FLORA OF TROPICAL EAST AFRICA

LOGANIACEAE

(including Buddleiaceae)

E. A. Bruce * and J. Lewis

Trees or shrubs (in our area ; also herbs elsewhere) with opposite simple leaves ; stipules interpetiolar, sometimes reduced to a mere line ; latex absent. Inflorescences usually cymose and moderately branched but sometimes of numerous cymules racemosely arranged. Flowers regular, usually hermaphrodite, sometimes heterostylous, usually 4–5-merous (corolla-lobes up to 20 in *Anthocleista*). Calyx shortly joined, ± campanulate. Corolla with a short or long tube ; lobes imbricate or valvate. Stamens equal in number to and alternate with the corolla-lobes (in our area ; more or less numerous elsewhere), the filaments often short. Disc absent or slight. Ovary superior or half-inferior, 2 (rarely 1 or 4)-locular ; style 1, rarely divided above ; stigmas 1, 2 or 4 ; ovules usually numerous, sometimes few (rarely 1) on axil placentas, anatropous or amphitropous. Fruit a usually septicidally dehiscent capsule (loculicidally dehiscent in *Mostuea*), berry or drupe. Seeds various, sometimes flattened or winged ; endosperm present ; embryo usually straight ; radicle usually inferior.

A tropical and warm-temperate world-wide family characterized by its opposite stipulate leaves, superior 2-locular ovary and lack of milky sap but otherwise very heterogenous. The genus *Gaertnera* Lam., included in this family in the "Flora of Tropical Africa," is now excluded ; otherwise the traditional conception of the family has been retained.**

Plants lacking glandular hairs ; flower parts in whorls of various numbers (4, 5 or 20) ; fruit a firm-walled berry, a drupe or a flattened capsule with two loculi containing one seed each ; intraxylary phloem present (subfamily **Loganioïdeae**) :
- Thin-stemmed shrubs, up to 4 m. tall, bearing usually solitary or few flowers ; corolla with 5 short ± rounded lobes ; fruit a laterally flattened capsule. 1. **Mostuea**
- Trees, thick-stemmed shrubs or climbers more than 5 m. tall bearing cymes of flowers ; corolla with either 10–20 oblong or 4–5 deltoid to oblong lobes ; fruit a berry or a drupe :
  - Leaves often more than 15 cm. long, glabrous ; corollas 1–2·5 cm. long, ± fleshy ; corolla-lobes more than 10 . . . . 2. **Anthocleista**

* Miss E. A. Bruce died on 13 Oct. 1955. This account was completed by Mr. Lewis before he left Kew in Feb. 1957.

** As generally understood, the *Loganiaceae* contains genera which individually show greater affinity with other families than among themselves. Affinities are to be found with *Apocynaceae*, *Scrophulariaceae*, *Gentianaceae* and *Solanaceae*, whilst *Gaertnera* Lam. and its allies have already been transferred to *Rubiaceae* (see p. 45). Dr. J. Hutchinson has decided to divide the *Loganiaceae* into five distinct families, often following the neglected opinions of earlier botanists ; he has published his new classification in the second edition of his "Families of Flowering Plants." He generously placed his manuscript at our disposal for this Flora, but it was decided not to make such a major change in this work, which deals with relatively few of the genera concerned.

Leaves usually less than 10 cm. long, often hairy ; corolla less than 1 cm. long, not fleshy ; corolla-lobes 4–5 . . . . 3. **Strychnos**

Plants possessing glandular hairs ; flowers 4-merous, rarely 5-merous ; fruit an obloid to globose capsule (sometimes with fleshy or fibrous walls) containing ± numerous seeds or rarely a thin-walled berry ; intraxylary phloem absent (subfamily **Buddleioïdeae**) :

Stem square in section and ± winged . . . . 5. **Adenoplusia**

Stem circular in section, not winged :

Abaxial leaf-surface concealed by a dense though sometimes thin covering of matted yellowish or brownish hairs ; calyx similarly hairy, not glandular-sticky ; stamens usually included within corolla-tube (exserted only in *B. dysophylla*) :

Fruit a capsule . . . . 4. **Buddleia**

Fruit a berry . . . . **Nicodemia** (pp. 35, 41)

Abaxial leaf-surface glabrous or visible through its sparse pubescence ; calyx glabrous, glandular-sticky ; stamens exserted . . . . . 6. **Nuxia**

1. MOSTUEA

Didr. in Kjoeb. Vidensk. Meddel. 1853 : 86 (1853)

Small much-branched shrubs. Leaves opposite, ± elliptic, entire. Inflorescences of few-flowered usually axillary cymes. Flowers heterostylous. Calyx regularly 5-lobed ; lobes longer than the tube. Corolla funnel-shaped, white, with or without yellow patches ; lobes 5, shorter than the tube. Stamens 5, alternate with the corolla lobes, inserted low in the corolla-tube. Ovary superior, 2-locular with 2 large erect basal ovules in each loculus ; styles 4 or fewer and then variously divided to support 4 stigmas. Capsule flattened at right-angles to the septum, forming an almost plane fruit broader than long with an emarginate apex, dehiscing loculicidally ; seeds 1–2 in each loculus.

A mainly African genus of which *Coinochlamys* T. Anders. (mostly from West Africa) might be considered a part, differing only in its possession of pairs of large thin bracts. Associated with the Asiatic and American genus *Gelsemium* Juss. to form the tribe *Gelsemieae*.

Most of the species of this genus are separable only partially and with difficulty, and this will remain so until, together with its allies, it undergoes detailed analysis. At present there are not sufficient data to decide which characters are specific. In the following account therefore only the more obvious synonyms have been eliminated.

In the circumstances it has been considered inappropriate to name new taxa on the basis of the unusual combinations of characters which have been encountered : two specimens have been placed as rather anomalous infraspecific variants and another three are given short diagnoses at specific level.

Branchlets ± regularly and ± densely hairy when young ; most leaves less than 4 cm. long, 2 cm. wide :

Plants more than 1 m. tall :

Abaxial leaf-surface irregularly and ± sparsely pubescent . . . . 1. *M. walleri*

Abaxial leaf-surface regularly, densely and shortly puberulous . . . . 5. *M.* sp. A

Plants up to 1 m. tall :

Calyx-lobes ± broadly elliptic or obovate, rounded (rarely subacute), glabrous or sparsely hairy above . . . . 4. *M. microphylla*

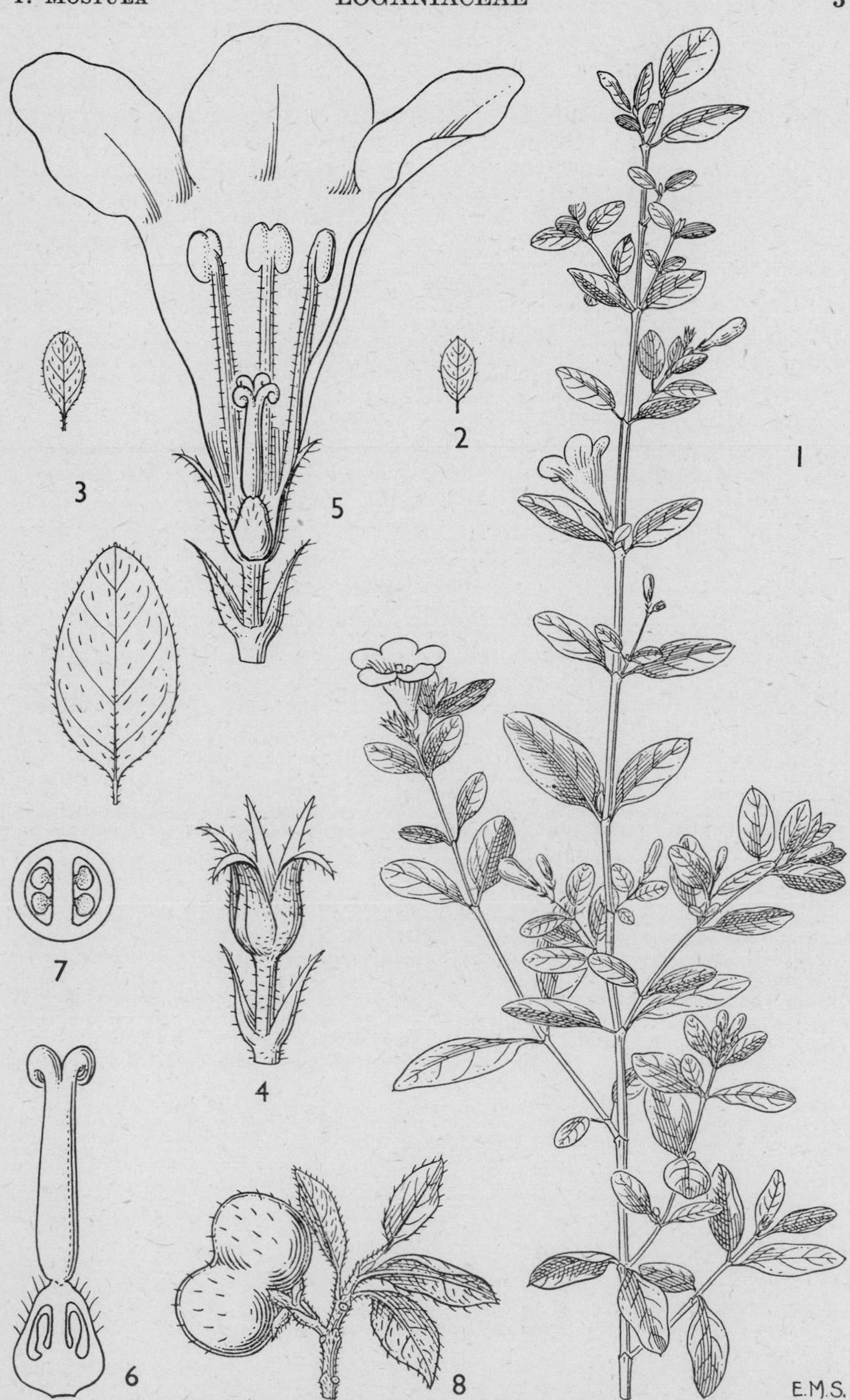

FIG. 1. *MOSTUEA WALLERI*—**1,** flowering branch, × ⅔ ; **2, 3,** leaves × 1 ; **4,** bracteoles, pedicel and calyx, × 4 ; **5,** flower, partly sectioned to show 3 stamens and pistil, × 4 ; **6,** pistil, with ovary sectioned to show position of ovules, × 8 ; **7,** T.S. of ovary (diagrammatic) ; **8,** fruiting branch, × 2. 1, from *Rounce* 509 ; 2, from *Drummond & Hemsley* 2812 ; 3, from *E. M. Bruce* 506 ; 4–8, from *Drummond & Hemsley* 2563.

Calyx lobes not shaped as above, distinctly acute, ± hirsute :
  Peduncles up to 12 mm. long, filiform, glabrous ; leaves up to 1·5 cm. long . . . . 6. *M*. sp. B
  Peduncles up to 2 mm. long, pubescent :
    Calyx-lobes lanceolate or ovate-lanceolate ; corolla-tube up to 10 mm. long ; leaves up to 5 cm. long . . . . . . 2. *M. camporum*
    Calyx-lobes narrowly triangular ; corolla-tube up to 6 mm. long ; leaves up to 1·5 cm. long . . . . . . 7. *M*. sp. C
Branchlets glabrous or with an irregular short sparse puberulence ; leaves more than 4 cm. long, 2 cm. wide (up to 8 cm. long, 4 cm. wide) . . . 3. *M. rubrinervis*

1. **M. walleri** *Baker* in K.B. 1895 : 96 (June* 1895) & in F.T.A. 4 (1) : 507 (1903) ; E. A. Bruce in K.B. 1956 : 159 (1956). Type : Portuguese East Africa, Zambezia, Mt. Morrumbala, *Waller* (K, holo. !)

Small, much-branched shrub, 1–5 m. high ; young branchlets with rather stiff, ascending, curved, white or brownish hairs, glabrate. Leaves petiolate ; petiole up to 3 mm. long ; lamina elliptic, ovate or obovate, 0·8–2(–3) cm. long, 0·4–1·2(–1·7) cm. wide, rounded to acute and apiculate, cuneate at the base ; midrib and lateral nerves beneath with angle-pockets and sparsely setose ; margin and upper surface sometimes with a few setae, otherwise glabrous. Inflorescence 1–3-flowered, terminal and axillary ; peduncle up to 2–3 cm. long ; pedicel 1–18 mm. long. Flowers heterostylous. Calyx-lobes linear, 3–5 mm. long, pilose or setose. Corolla-tube 7–14 mm. long, white with yellow within the base of the tube. Stamens inserted near the base of the tube ; filaments shortly pubescent ; anthers of thrum-eyed flowers reaching to the throat or shortly exserted. Fruit obreniform, compressed, about 12 mm. wide, ± pubescent or setose. Seeds orbicular, flattened on one side.

KENYA. Teita District : Sagalla Mt., 4 Feb. 1953 (fl. & fr.), *Bally* 8710 ! ; Kilifi District : Rabai Hills, Mwachewatini, *W. E. Taylor* !
TANGANYIKA. Lushoto District : W. Usambara Mts., Dolelo, 4 km. NW. of Kwai, 29 May 1953 (fl.), *Drummond & Hemsley* 2812 ! & Shagai Forest, 15 May 1953 (fl. & fr.), *Drummond & Hemsley* 2563 ! ; Morogoro, Morningside, 5 Dec. 1934 (fl.), *E. M. Bruce* 252 !
DISTR. **K**7 ; **T**3, 4, 6, 8 ; southwards to Nyasaland, Portuguese East Africa & eastern Southern Rhodesia.
HAB. Upland & lowland rain-forest ; 100–2200 m.

SYN. *M. grandiflora* Gilg in P.O.A. C : 310 (Aug. 1895) ; F.T.A. 4 (1) : 507 (1903) ; T.T.C.L. : 272 (1949). Types : Tanganyika, Usambara Mts., Lutindi, *Holst* 3430 (B, syn.†, H, K, isosyn. !) & Magamba, *Holst* 3813 (B, syn.†)
M. ulugurensis Gilg in E.J. 23 : 198 (1896) ; F.T.A. 4 (1) : 506 (1903) ; T.T.C.L. : 272 (1949). Type : Tanganyika, Uluguru Mts., Nglewenu, *Stuhlmann* 8865 (B, holo.†, K, fragment !)
M. gracilipes Mildbr. in N.B.G.B. 11 : 675 (1932) ; T.T.C.L. : 272 (1949). Type : Tanganyika, Ulanga District, near Massagati, *Schlieben* 1115 (B, holo.†, BR, iso. !)

NOTE. The inclusion of the last synonym is debatable ; the calyx-lobes are sometimes subspathulate (like those of *sp. B* below) and the inflorescences are unusually longly stalked for *M. walleri*. However, total comparison supports the relationship of *Schlieben* 1115 and *M. walleri* being closer than that between either of them and *Eggeling* 6418 (*sp. B*). This is confirmed by *Schlieben* 5935 from Lindi District, Lake Lutamba (BR !) which although certainly conspecific with *M. gracilipes*, is even more like *M. walleri* than is *Schlieben* 1115.
See also *M. erythrophylla* Gilg (p. 7), which may be another synonym for this species.

* See Britten in J.B. 33 : 255 (1895).

VARIATION. Two of the variants mentioned in the note on the genus (p. 2) have been provisionally determined as belonging to this species. One is *Greenway* 2898 from Tanganyika, E. Usambara Mts., Mangubu–Misoswe ; an uncommon shrub, 1 m. tall, in the shade of very open rain-forest on a mountain slope at 400 m. Although in many ways like *M. walleri*, it differs slightly with indumentum less appressed, the flowers smaller than usual and the leaves relatively broader and more often apically rounded. If it is this species, the congested nature of the inflorescences and the basally broader calyx-lobes make it a very distinct form.

The other is *Drummond & Hemsley* 3788 from Kenya, Kwale District, between Umba and Mwera rivers on the Lungalunga–Msambweni road, described as a shrub 2 m. tall in coastal forest at 100 m. The indumentum, flower colour, shape and size of the leaves all associate this specimen with our species. The unequal calyx-lobes, the shorter corolla and, especially, the densely pilose nature of the fruit (which also differs in its detailed shape) make it a very anomalous example of it.

2. **M. camporum** *Gilg* in E.J. 28 : 117 (1899) ; F.T.A. 4 (1) : 509 (1903) ; T.T.C.L. : 272 (1949). Type : Tanganyika, Kilosa District, Khutu Plain, Mgunda, *Goetze* 379 (B, holo.†, BM, K, iso. !)

Small shrub, 0·3–0·6 m. tall ; branchlets ± densely and shortly rough-pubescent, glabrate. Leaves shortly petiolate ; lamina ovate-lanceolate to broadly ovate, 1·2–2 cm. long, 0·5–1·1 cm. wide, acute or obtuse, narrowly to widely cuneate at the base ; midrib and lateral nerves beneath thinly pubescent, glabrescent ; upper surface and margins very sparsely longly pubescent. Inflorescence 1–2(–3)-flowered, terminal (and on short lateral branches, appearing axillary) ; peduncle absent or up to 2 mm. long, pedicels 1–2 mm. long, pubescent. Flowers heterostylous. Calyx-lobes lanceolate to ovate-lanceolate, 2·5–3 mm. long, long-acuminate, ciliate. Corolla-tube 8–10 mm. long, white. Stamens inserted near the base of the tube ; filaments of pin-eyed flowers long, shortly pubescent, about half the length of the tube (and style) ; in thrum-eyed flowers the reverse. Ovary ovoid, with a few apical hairs. Fruit not known.

TANGANYIKA. Morogoro District : Morogoro–Mgeta road, Mbingu, 28 Sept. 1952 (fl.), *Carmichael* 129 ! ; Njombe District : Lupembe, Oct. 1931 (fl.), *Schlieben* 1345 ! ; Lindi District : Nachingwea, 10 Nov. 1951 (fl.), *Evans* 27 !

DISTR. T6–8 ; not known elsewhere.

HAB. Deciduous woodland, on red sandy loam ; 400–1600 m.

VARIATION. *Schlieben* 5467, from Tanganyika, Lindi District, Lake Lutamba, is very like *M. camporum* in its twiggy branches and short stature (20–80 cm.) but differs in the smaller flowers and appressed-pubescent calyx-lobes.

3. **M. rubrinervis** *Engl.* in E.J. 7 : 340 (1886) ; F.T.A. 4 (1) : 508 (1903). Type : Kenya, Mombasa, *Wakefield* (B, holo.†, K, iso. !)

Small shrub, 0·7–2 m. high ; branchlets glabrous or irregularly shortly and sparsely puberulent. Leaves shortly petiolate ; lamina ovate, ovate-lanceolate or elliptic, 2·5–10 cm. long, 1–4 cm. wide, acuminate or narrowed to the apex then usually rounded and apiculate, more rarely acute, cuneate at the base, glabrous on both surfaces, except for angle-pockets beneath. Cymes usually (2–)3-flowered, terminating short lateral branches ; peduncle 1–2·5 cm. long ; pedicels 0·2–1·0 cm. long, usually glabrous. Calyx-lobes scarcely 2 mm. long, equal, ovate-lanceolate, usually acuminate, coarsely ciliate, otherwise glabrous. Corolla white, with a yellow area near the base of the 6–11 mm. long tube. Filaments about 4 mm. long ; anthers about 1·5 mm. Ovary ovoid, glabrous. Fruit obreniform, compressed, glabrous, about 12 mm. wide. Seeds 2 in each loculus, quadrangular-orbicular, ± concave on one side, about 5 mm. in diameter, finely pitted or reticulate, brown-pubescent.

KENYA. Kwale District : Buda Mafisini Forest, 13 km. WSW. of Gazi, Aug. 1935 (fl. & fr.), *Drummond & Hemsley* 3805 ! ; Kilifi District : Sokoke, *R. M. Graham* 1949 ! ; Tana River District : Gongoni Forest [NW. of Witu], *Dale* 3809 !

TANGANYIKA. Morogoro District: Turiani, by R. Liwali, Mar. 1953 (fl. & fr.), *Drummond & Hemsley* 1798 & 1806; Uzaramo District: Vikindu Forest Reserve, Aug. 1953 (fl.), *Paulo* 136! & Pugu Hills, Jan. 1937 (fr.), *Maclean* 4!
DISTR. **K**7; **T**3, 6; not known elsewhere.
HAB. Lowland rain-forest & riverine forest; up to 800 m.

SYN. *M. orientalis* Baker in K.B. 1895: 96 (1895); T.S.K.: 124 (1936). Type: as *M. rubrinervis* (K, holo.!)

NOTE. *M. schumanniana* Gilg, a species from French Equatorial Africa of which an isotype from near Mayumbe, in French Congo (*Soyaux* 136), is at Kew, is scarcely specifically distinct from this species. Such affinity as the few South American species of the genus have with those of Africa is with *M. rubrinervis*.

4. **M. microphylla** *Gilg* in P.O.A. C: 310 (1895); F.T.A. 4 (1): 505 (1903); T.T.C.L.: 272 (1949); E. A. Bruce in K.B. 1956: 160 (1956). Type: Tanganyika, Uzaramo District, Pugu Hills, Kiserawe, *Stuhlmann* 6172 (B, holo.†)

Small shrub, 0·3–1 m. tall; branchlets sparsely pubescent or hispid, glabrate. Leaves shortly petiolate, ovate to ovate-lanceolate, 1–3·2 cm. long, 0·4–1·5 cm. wide, rounded, subacute, emarginate, narrowly to widely cuneate or rounded at the base, glabrous on both surfaces except for a few hairs along the midrib above and hair pockets in the angles beneath. Inflorescence (2–)3(–4)-flowered cymes, axillary or terminal on short branches; peduncle 0·3–2 cm. long; pedicels 0·3–0·8 cm., like the peduncle slender and glabrous or very sparsely pubescent. Calyx-lobes elliptic, about 1·5 mm. long, obtuse or subacute, sometimes very shortly and sparsely ciliate otherwise glabrous. Corolla white, the tube 7–9 mm. long. Filaments puberulous, slender, 4–5 mm. long in thrum-eyed flowers, much exceeding the styles. Ovary narrowly ovoid. Fruit not known.

KENYA. Kilifi District: Rabai Hills, 10 Nov. 1885, *W. E. Taylor*!; Lamu District: Kitangani, 27 Dec. 1946 (fl.), *Bally* 5824!
TANGANYIKA. Uzaramo District: Pugu Hills [Sachsenwald], *Holtz* 870!; Lindi Creek, 12 Dec. 1942 (fl.), *Gillman* 1167!; 20 km. SW. of Lindi, Mlingura, 18 Dec. 1934 (fl.), *Schlieben* 5747!
DISTR. **K**7; **T**6, 8; coast of Portuguese East Africa.
HAB. Coastal evergreen bushland & lowland rain-forest; up to 350 m.

SYN. *M. syringiflora* S. Moore in J.B. 44: 24 (1906). Type: Kenya, Kilifi District, Rabai Hills, *W. E. Taylor* (BM, holo.!)
M. amabilis Turrill in K.B. 1920: 25 (1920). Type: Portuguese East Africa, Msalu River, *Allen* 90 (K, holo.!)

Imperfectly known species

5. **M. sp. A.**

Shrub, 1·8 m. tall; very young parts densely white-pubescent; branchlets puberulous. Leaves shortly petiolate; blade elliptic, about 2 cm. long, 1 cm. wide, narrowing above to a blunt apex, drying dark green, puberulent adaxially. Inflorescence few-flowered, terminal, congested. Calyx-lobes oblong, acute, ± 3 mm. long, exceeding the greenish-cream corolla in the opening bud; the whole young flower densely and regularly puberulous. Fruit not known.

TANGANYIKA. Lindi District: S. face of Rondo escarpment at Mchinjiri, Dec. 1951 (fl.), *Eggeling* 6419!
DISTR. **T**8; not known elsewhere.
HAB. Not known; 780 m.

NOTE. Although it is allied to *M. walleri*, this plant is unique in its indumentum. All other members of the genus have the upper leaf-surface either glabrous or with few irregularly distributed rather long hairs mostly on the midrib. *Eggeling* 6419 has leaves perfectly regularly covered above with hairs no longer than twice their thickness. On the calyx-lobes the indumentum is longer, but still very regular, and this together with the distinctly oblong shape of these lobes and the congestion of the inflorescence, gives the plant a most unusual appearance.

6. **M. sp. B.**

Shrub 0·7 m. tall ; branchlets pubescent with brownish curved-ascending hairs. Leaves shortly petiolate ; blade elliptic, about 1·2 cm. long, 0·6 cm. wide, narrowing to a subtruncate and apiculate apex. Inflorescence of solitary or, more rarely, paired flowers on long slender glabrous pedicels up to 1·5 cm. long. Calyx-lobes oblong or narrowly oblanceolate-spathulate, acute, pubescent. Corolla white, yellow within, about 8 mm. long. Fruit not known.

Tanganyika. Lindi District : S. face of Rondo escarpment at Mchinjiri, Dec. 1951 (fl.), *Eggeling* 6418 !
Distr. **T8** ; not known elsewhere.
Hab. Not known ; 780 m.

Note. This specimen is also allied to *M. walleri* ; the habit, the few flowers, the type of indumentum, the leaf-shape and -size—all these could be accepted in that species. However, the exact shape of the calyx-lobes, their shorter indumentum, the smaller flowers and especially the very long pedicels make this plant recognizably distinct from any of the above species.

7. **M. sp. C.**

Shrub up to 0·6 m. tall ; branchlets with rather stiff curved white or brownish hairs, glabrate. Leaves shortly petiolate ; lamina elliptic or ovate, up to 1·6 cm. long, 0·8 cm. wide, narrowing above to a blunt and sometimes apiculate apex, setose on the midrib and nerves beneath. Inflorescence 1–3-flowered ; peduncle and pedicels very short. Calyx-lobes narrowly triangular, acute, shortly and sometimes sparsely setose. Corolla-tube up to 6 mm. long. Fruit obcordiform, compressed, about 0·9 cm. wide, puberulous and setose.

Tanganyika. Pangani District : Madanga, 28 May 1957 (fl.), *Tanner* 3522 ! & 24 July 1957 (fl. & fr.), *Tanner* 3626 !
Distr. **T3** ; not known elsewhere.
Hab. Dry evergreen forest and regenerating abandoned cultivations ; up to 75 m.

These two specimens might be considered as variants of *M. walleri* of very short stature having smaller leaves and flowers. The smaller fruit of different shape and with a double indumentum however makes it preferable to treat it separately.

Doubtful species

M. erythrophylla *Gilg* in E.J. 28 : 117 (1899) ; T.T.C.L. : 272 (1949). Type : ? Tanganyika or Urundi, E. shore of Lake Tanganyika, *v. Trotha* 10 (B, holo.†)

Shrub, 2 m. tall ; branchlets densely yellow-pubescent. Leaves ± ovate, up to 2·3 cm. long, 1·2 cm. wide, rounded above, glabrous. Inflorescences terminal, lax, 3–7-flowered ; peduncles about 0·5 cm. long. Calyx-teeth ovate-lanceolate, acute, sparsely pilose. Corolla 1–1·5 cm. long.

? Tanganyika. E. shore of Lake Tanganyika, *v. Trotha* 10
Distr. ? **T4** or possibly Urundi ; not known elsewhere.
Hab. Not known ; about 760 m.

Allied to *M. walleri* ; Gilg himself suggests an affinity with *M. ulugurensis*. The type-specimen has not been examined, but the description shows a number of unusual features.

2. ANTHOCLEISTA

[Afzel. ex] R. Br. in Tuckey, Congo, App. : 449 (1818) ; E.J. 17 : 575 (1893) ; E. A. Bruce in K.B. 1955 : 45 (1955)

Erect trees with a usually straight unbranched trunk below and candelabra branching above or, more rarely, scandent shrubs. Branchlets sometimes armed with short paired spines ; leaf-scars ± sickle-shaped, often prominent on the branchlets and cohering in opposite and decussate pairs. Leaves in the tree-species often large and grouped at the ends of the branches, giving a cabbage-like appearance, but not so in scandent species, sessile or petiolate, sometimes auriculate. Cymes terminal, usually trichotomous and compound. Flowers large, usually sweet-scented. Calyx-lobes 4, opposite and decussate, coriaceous or fleshy, usually much imbricated. Corolla white, cream or flesh-coloured, browning with age ; tube cylindrical to funnel-shaped ; lobes 6–20, contorted and usually much overlapping. Stamens as many as the corolla-lobes ; filaments united into a collar at the base, adhering to the corolla-tube ; anthers subsessile. Ovary globose to ovoid, spuriously 4-locular, with ∞ ovules inserted on 2–4 fleshy placentas ; disc annular, fleshy ; style usually subequal to the corolla-tube ; stigma large, capitate or turbinate. Fruit a globose, ovoid or broadly fusiform, firm-walled berry. Seeds numerous, embedded in fleshy pulp.

A genus of the tribe *Fagraeeae*, confined to Africa, Madagascar and the Comoro Is., which is most clearly characterized by its corolla being contorted sinistrorsely in aestivation.

Branches armed with short paired spines ; corolla-tube not much longer than the calyx and subequal to the corolla-lobes ; flower-buds uniformly rounded at the apex . . . . 1. *A. vogelii*

Branches unarmed ; corolla-tube much longer than the calyx and longer than the corolla-lobes ; flower-buds usually tapered to the apex :

Leaves sessile or subsessile, usually membranous, with conspicuous tertiary nerves ; calyx-lobes rugulose, outer pair at least ± spreading, not closely clasping the corolla-tube during flowering, widely spreading in fruit ; fruits narrowly ovoid, often umbonate at the apex, at least when young . . . . . . 2. *A. zambesiaca*

Leaves, at least the upper ones, petiolate, lower ones often subsessile, usually subcoriaceous, with indistinct tertiary nerves ; calyx-lobes not rugulose, clasping the corolla-tube during flowering, sometimes spreading in mature fruit ; fruits broadly ovoid to globose, rounded above 3. *A. schweinfurthii*

1. **A. vogelii** *Planch.* in Hook., Ic. Pl. tt. 793, 794 (1848) ; F.W.T.A. 2 : 18, fig. 184 (1931) ; Fl. Pl. Sudan 2 : 380, fig. 142 (1952) ; E. A. Bruce in K.B. 1955 : 48 (1955). Type : Nigeria, Ibu [Ibo], *Vogel* 51 (K, holo. !)

Tree 5–18 m. high, with smooth light-grey bark, sparsely branched at the top with a spreading crown ; spines in pairs on the branchlets, arising well above the leaf-base, permanent on the stems of old trees and becoming woody. Leaves normally sessile, rarely shortly petiolate with petiole up to 2·5 cm. long ; lamina coriaceous, generally obovate, sometimes elongate-obovate or oblanceolate, 15–50 cm. long, 8–24 cm. wide (the upper leaves

FIG. 2. *ANTHOCLEISTA ZAMBESIACA*—**1,** apex of leafy branch, × ½ ; **2,** inflorescence, × ⅔ ; **3,** corolla partly sectioned to show stamens, × 1 ; **4,** pistil, × 1 ; **5,** fruit, × ⅔. 1, from *Drummond & Hemsley* 1672 ; 2, from *Verdcourt* 225 ; 3–5, from *Drummond & Hemsley* 3458.

smallest), rounded to broadly rounded at the apex, cuneate or narrowly cuneate, auriculate at the base ; lateral nerves 11–15, prominent beneath ; tertiary nerves not or rarely visible. Inflorescence shorter than the subtending leaves, 20–24(–40) cm. long ; primary peduncle 7–20(–26) cm. long, secondary 3–6(–8) cm. long ; pedicels very short and stout ; bracts leathery, triangular, acute. Outer calyx-lobes leathery, orbicular or subquadrate, 1–1·2 cm. long, about 1 cm. broad, rounded or emarginate ; inner lobes slightly longer and broader, overlapping each other, up to 1·5 cm. broad, truncate at the apex, clasping the corolla-tube. Corolla-tube 1–2 cm. long ; corolla-lobes 13–15, 1–2 cm. long, oblique-oblong, often cohering at the apex and not fully opening, in bud always rounded and blunt at the apex and only a little longer than the calyx in the early stage but becoming longer with age. Anthers about 1 cm. long ; staminal-tube 3–4 mm. long. Ovary ovoid, about 6 mm. long ; style about 1·5 cm. long ; stigma turbinate, longer than broad, about 5 mm. long. Fruits generally erect, globose or ovoid, 2·5–4 cm. long, 1·8–3 cm. in diameter, clasped by the calyx.

UGANDA. West Nile District : Logiri, Mar. 1935 (fl.), *Eggeling* 1858 ! ; Ankole District : Ruampara, Rwoho, 7 Sept. 1949 (fl.), *St. Clair-Thompson* 1815 ! ; Masaka District : Buddu, *Fyffe* 14 !

KENYA. N. Kavirondo District : Kakamega, June 1950, *Feltham in E.A.H.* 10290 !

TANGANYIKA. Bukoba District : 19 km. W. of Bukoba, July 1951 (fl. & fr.), *Eggeling* 6251 ! ; Buha District : between Kibondo & Mabamba, Nov. 1956, *Procter* 593

DISTR. **U**1, 2, 4 ; **K**5 ; **T**1, 4 ; Belgian Congo, Cabinda, the Sudan, Northern Rhodesia and West Africa from the Cameroons to Sierra Leone.

HAB. Swamp-forest, both primary and secondary ; 1200–1400 m.

SYN. [*A. nobilis* sensu Baker in F.T.A. 4 (1) : 538, quoad spec. ex Nigeria, Ibu, *Vogel* 51 ! ; I.T.U., ed. 2 : 164 (1952) ; ? T.T.C.L. : 269 (1949) ; *non* G. Don]

NOTE. A note on *Bagshawe* 1459 (from Uganda, Bunyoro, near Hoina) states that the trunk is unarmed, though the specimen appears otherwise to be typical of *A. vogelii*.

Five 'species' from outside our area are included within this species : *A. kalbreyeri* Baker, *A. zenkeri* Gilg and *A. talbotii* Wernh. from West Africa ; *A. auriculata* De Wild. and *A. bequaertii* De Wild. from Belgian Congo.

2. **A. zambesiaca** *Baker* in K.B. 1895 : 99 (1895) ; P.O.A. C : 424 (1895) ; F.T.A. 4 (1) : 540 (1903) ; Prain & Cummins in Fl. Cap. 4 (1) : 1049 (1909) ; U.O.P.Z. : 125 (1949) ; E. A. Bruce in K.B. 1955 : 54 (1955). Type : Nyasaland, Shire Highlands, *Buchanan* 84 (K, holo. !)

Tree 7–27 m. high ; bark transversely striate ; branchlets unarmed. Leaves sessile or rarely shortly petiolate with petiole up to 2 cm. long, very variable in size ; lamina usually membranous, sometimes more or less leathery, oblanceolate to broadly obovate, 20–60(–120) cm. long, 8–30(–50) cm. wide, rounded or subacute, cuneate at the base, not or inconspicuously auriculate ; margin more or less inrolled and crinkled ; lateral nerves 9–14, prominent below ; tertiary nerves conspicuous. Inflorescence 19–40 cm. long, 10–25 cm. broad, usually erect and shorter than the subtending leaves ; primary peduncles up to 25 cm. long ; pedicels fairly stout, 4–10 mm. long ; bracts leathery, broadly ovate-deltoid, acute, 4–8 mm. long. Flowers fragrant. Outer calyx-lobes orbicular, rather fleshy or leathery, rugose, 6–10 mm. long, rounded to subacute, often rather spreading and not clasping the corolla-tube ; inner lobes subequal to outer ones, sometimes broader, less leathery and rugose, subacute, rounded or rarely emarginate, not or scarcely overlapping each other. Corolla-tube 2·5–3 cm. long ; corolla-lobes (10–)12(–16), oblong or very narrowly elliptic, 1·3–2·2 cm. long, often ± recurved ; buds usually tapered and definitely broadest below the apex. Anthers 6–7 mm. long. Ovary ovoid to narrowly ovoid, 5–9 mm. long ; style 2·5–4 cm. long ; stigma subglobose, 2–3 mm. in diameter, just exserted in the open flower. Fruit erect, more or less ovoid, 3–3·5 cm. long, 2–2·5 cm.

in diameter, usually beaked or pointed at the apex, wrinkled with the stalk thickened, and the calyx woody, spreading, the lobes shrivelling and becoming subacute.

Uganda. Ruwenzori, Mubuku Valley, Aug. 1933 (fl.), *Eggeling* 1257 ! ; Mbale District : W. of Elgon, Ngongoro Ridge, Nov. 1939 (young fr.), *Dale* U78 !

Kenya. Meru/Embu District : near Mutonga River, 24 Feb. 1922 (young fl.), *Fries* 1902 ! ; Kericho District : Sotik, *Battiscombe* 1301 ! & SW. Mau Forest Reserve, 13 Aug. 1949 (young fr.), *Maas Geesteranus* 5751 !

Tanganyika. Lushoto District : E. Usambara Mts., Amani, Kwamkoro road, 27 May 1950 (fl.), *Verdcourt* 225 ! & Kwamkoro, 25 July 1953 (fl. & fr.), *Drummond & Hemsley* 3458 ! ; Rungwe District : Kyimbila, 31 Dec. 1910, *Stolz* 497 !

Zanzibar. Zanzibar Is., Jozani Forest, 20 Dec. 1929 (fl.), *Vaughan* 997 !

Distr. **U**1–4 ; **K**4, 5 ; **T**3, 6, 7 ; **Z** ; Belgian Congo, Portuguese East Africa, Nyasaland, Southern Rhodesia, the Transvaal & Swaziland.

Hab. Upland & lowland rain-forest, particularly in swampy ground, riverine forest, often persisting where forest is destroyed ; 0–2300 m.

Syn. *A. orientalis* Gilg in P.O.A. C : 312 (Aug. 1895) ; F.T.A. 4 (1) : 539 (1903) ; Z.A.E. : 652 (1914) ; F.P.N.A. 2 : 63 (1947) ; T.T.C.L. : 269 (1949). Type : Tanganyika, Uzaramo District, Vikindu, *Stuhlmann* 6092 (B, holo.†)

A. pulcherrima Gilg in E.J. 30 : 374, t. 17 (1901) ; F.T.A. 4 (1) : 540 (1903) ; T.T.C.L. : 269 (1949) ; I.T.U., ed. 2 : 165 (1952). Type : Tanganyika, Rungwe District, Umuamba & Yungururu Crater Lake, *Goetze* 1313 (B, holo.†)

A. keniensis Summerh. in K.B. 1926 : 244 (1926) ; T.S.K. : 127 (1936). Type : Kenya, Kericho District, Sotik, *Battiscombe* 1301 (K, holo. !)

Variation. There is a certain amount of variation within the species in leaf-texture, size of calyx, number of corolla-lobes and stamens and in the stoutness of the inflorescence branches.

Note. The Swaziland species, *A. insignis* Galpin, is also synonymous.

3. **A. schweinfurthii** *Gilg* in E.J. 17 : 579 (1893) ; P.O.A. C : 312 (1895) ; F.T.A. 4 (1) : 541 (1903) ; T.T.C.L. : 269 (1949) ; I.T.U., ed. 2 : 165 (1952) ; E. A. Bruce in K.B. 1955 : 51 (1955). Types : Belgian Congo, Uele Province, R. Mbrwole, *Schweinfurth* 3726 (B, syn.†, K, isosyn. !) & Tanganyika, Bukoba, *Stuhlmann* 3751 (B, syn.†) & Mwanza District, Maisome Is., *Stuhlmann* 4133 (B, syn.†)

Tree 6–20 m. high ; branchlets unarmed ; internodes 0·5–2(–3) cm. long. Leaves bright green, glossy, upper ones petiolate, lower ones often subsessile ; petiole 1·9 cm. long ; lamina usually subcoriaceous, oblanceolate or narrowly obovate, 13–45(–80) cm. long, 6–14(–23) cm. wide, acute, subacute or rounded, narrowly cuneate and often auriculate at the base ; auricles semi-orbicular, undulate, often drying up ; leaf-margin slightly undulate-crenulate ; lateral nerves 6–13, prominent below ; tertiary nerves often visible but not conspicuous. Inflorescence shorter than the subtending leaves, 8–24 cm. long ; primary peduncles 5–18 cm. long ; secondary 3–7 cm. long ; all comparatively slender ; pedicels ± slender, 1–1·5 cm. long (increasing to 2·5 cm. in fruit) ; bracts ovate to subquadrate, subacute, often conspicuous and persistent in fruit, about 8 mm. long. Flowers fragrant. Calyx-lobes not fleshy ; outer lobes suborbicular or subquadrate, 8–12 mm. long, rounded or truncate, inner slightly longer and broader, overlapping each other and clasping the corolla-tube. Corolla-tube 2·4–2·8 cm. long ; corolla-lobes about 14, ± oblong, 1·6–2·2 cm. long, rounded and suboblique, often reflexed ; buds tapered or pointed when young, becoming less acute but always broader below the apex. Fruit ovoid to globose, 2·5–3·6 cm. long, 2–3 cm. in diameter apically often apiculate (remains of style-base) or with a small terminal hole, ± clasped at the base by the calyx.

Uganda. Masaka District : Sesse Islands, Bugala, near Sozi Point, Nov. 1931 (fr.), *Eggeling* 89 *in F.D.* 262 ! & *Philip* 433 ! & Katera, 1 Oct. 1953 (fr.), *Drummond & Hemsley* 4504 !

TANGANYIKA. Buha District : Kasulu–Kibondo road, R. Mlagarasi, *Bullock* 3223 !
DISTR. **U**2, 4 ; **T**1, 4 ; Nigeria, Cameroons, Gabon, Cabinda, Belgian Congo and the Sudan.
HAB. Lowland rain-forest, often persisting in secondary evergreen bushland ; 1200–1400 m.

SYN. *A. stuhlmannii* Gilg in E.J. 17 : 580 (1893) ; P.O.A. C : 312 (1895) ; F.T.A. 4 (1) : 540 (1903) ; Z.A.E. : 533 (1913) ; T.T.C.L. : 269 (1949). Type : Tanganyika, Bukoba, *Stuhlmann* 3727 (B, holo.†)
[*A. nobilis* sensu Baker in F.T.A. 4 (1) : 538 (1903), quoad spec. *Schweinfurth* 3037 !, *non* G. Don.]
[*A. inermis* sensu Baker in F.T.A. 4 (1) : 541 (1903), quoad spec. ex Lower Congo, *Ch. Smith* ; I.T.U., ed. 2 : 164 (1952) ; *non* Engl.]
A. insulana S. Moore in J.L.S. 37 : 186 (1905) ; I.T.U., ed. 2 : 164 (1952). Type : Uganda, Mengo District, Buvuma Is., *Bagshawe* 638 (BM, holo. !)

NOTE. The following synonyms have been used for this species in other areas of Africa : *A. niamniamensis* Gilg, *A. magnifica* Gilg, *A. squamata* De Wild. & Th. Dur., *A. laurentii* De Wild., *A. pynaertii* De Wild., *A. oubanguiensis* Aubrév. & Pellegr. and perhaps *A. gigantea* Gilg.

3. STRYCHNOS

L., Sp. Pl. : 179 (1753) & Gen. Pl., ed. 5 : 86 (1754) ; Solered. in E. & P. Pf. 4 (2) : 37 (1892) ; Duvign. in B.S.B.B. 85 : 9 (1952)

Trees or shrubs, sometimes scrambling or climbing with spines or hooked tendrils. Leaves decussate, entire, 3–7-nerved from near the base ; stipules absent or represented by an interpetiolar ridge. Cymes axillary or terminal, simple or panicled. Calyx 4- or 5-lobed, sometimes very deeply. Corolla campanulate or salver-shaped ; lobes 4 or 5, valvate in the bud. Stamens 4 or 5 ; filaments short but anthers usually exserted. Ovary 2-locular ; style moderately long, undivided ; stigma usually capitate ; ovules few to many. Fruit a drupe or a berry with a hard rind. Seeds few to numerous, flattened or globose ; endosperm copious and hard ; embryo small.

A genus of numerous species widespread in tropical and subtropical areas of the world. It is of considerable economic interest.

Careful collecting and observing would enable many improvements in the taxonomy to be made ; the variations in habit need study ; the fruits are not at all well known and confirmation is needed of the positions of the inflorescences, whether axillary or terminal or both.

Leaves all about 5 mm. wide ; small tree without spines or tendrils . . . . 1. *S. myrtoïdes*
Leaves more than 10 mm. wide or, if less, then plant scandent and bearing tendrils :
  Leaves bearing a hard sharp apical spine . . . 9. *S. pungens*
  Leaves not spinous ; apex truncate to acuminate, and sometimes mucronulate :
    Small trees, shrubs or rarely climbers, usually bearing (paired) spines, at least on the older wood, and very occasionally with tendrils also ; calyx-lobes linear to narrowly triangular :
      Climber, scandent shrub or tree with pendulous branches ; all leaves rounded above to a distinctly acuminate apex . . . 2. *S. congolana*
      Erect shrubs or small trees ; leaves variable, usually narrowing to an acute or mucronulate apex or, if rounded above, then the apex acute or rarely acuminate :
        Bark of older branches deeply grooved into corky ridges ; calyx-lobes uniformly pubescent outside . . . . . 3. *S. cocculoïdes*

Bark of older branches shallowly fissured and powdery ; calyx-lobes glabrous above outside . . . . . . 4. *S. spinosa*

Trees, shrubs and climbers without true spines (one species having prickles and some bearing tendrils) ; calyx-lobes oblong, deltoid, elliptic, ovate, obovate or oblate :

Scandent shrubs, climbers or small trees with sarmentose or pendulous branches ; tendrils present* :

Lobes of the corolla shorter than the tube ; stamens usually inserted below the sinuses of the corolla (well within the tube or at the throat), not exserted (slightly so only in 8, *S. lucens*) :

Branches angular when young and bearing prickles on the angles ; fruits 9–12 cm. in diameter . . . . . . 5. *S. aculeata*

Branches not prickly :

Leaves ± papery, 3-nerved from about 1 cm. above the base ; tendrils glabrous ; stamens inserted at the base of the corolla-tube ; filaments as long as the anthers ; fruit about 7 cm. in diameter . . . 2. *S. congolana*

Leaves coriaceous, 3-nerved at or near the base ; tendrils pubescent ; stamens inserted well above the base of the corolla-tube ; anthers sessile ; fruit 1–5 cm. in diameter :

Scandent shrub or woody climber, about 4 m. tall ; tendrils simple ; flowers pale yellow ; leaves ± $3{\cdot}5 \times 2{\cdot}5$ cm. . . . . 6. *S. matopensis*

Small tree with sarmentose branches, semi-scandent shrub or woody climber, 5–20 m. tall ; tendrils bifurcate ; flowers greenish-white ; leaves $5 \times 3{\cdot}5$ cm. or larger (Fig. 4, p. 22) . . 8. *S. lucens*

Lobes of the corolla equal to or longer than the tube ; stamens inserted at the sinuses and exserted (only slightly so in 13, *S. angolensis*) :

Tendrils bifurcate ; buds up to 6 mm. long ; fruits hard-walled ; seeds (1–)3–15 :

Leaves up to 2·5 cm. wide :

Young branchlets velutinous ; cymes congested, up to 1·5 cm. long ; fruits up to 4 cm. in diameter with up to 30 seeds . . . 16. *S. kasengaënsis*

* Usually, on the plant, not necessarily on a gathered specimen. In case of doubt at this division of the key, try the other alternative route first ; a scandent plant will usually be easily recognizable.

Young branchlets pubescent ; cymes lax, up to 2·5 cm. long ; fruits up to 1·6 cm. in diameter with up to 5(–10) seeds . . . 12. *S. panganensis*

Leaves 3 cm. wide or more ; young branchlets glabrous or sparsely pubescent ; cymes somewhat congested, up to 2·0 cm. long ; fruits up to 3·5 cm. in diameter with up to 8(–15) seeds (Fig. 4, p. 22) . 17. *S. scheffleri*

Tendrils simple ; buds up to 3 mm. long ; fruits * soft-skinned ; seeds 1(–2) :

Branchlets glabrous ; leaves narrowly ovate, glabrous ; cymes predominantly axillary, less often terminal ; ovary glabrous . . . . 21. *S. usambarensis*

Branchlets hairy ; leaves ovate, elliptic or obovate with hairy midrib ; cymes variously arranged ; ovary glabrous or pilose :

Cymes densely congested, seldom axillary ; leaves not cuneate at the base, acute, sometimes slightly acuminate ; petiole up to 3 mm. long ; ovary glabrous ; young branchlets velutinous . 6. *S. matopensis*

Cymes not or little congested, often axillary ; petiole up to 5 mm. long:

Leaves cuneate at the base, narrowly acuminate with an acute apex, membranous ; ovary pilose ; young branchlets irregularly puberulous . . 15. *S. boonei*

Leaves not cuneate at the base, round, emarginate, or rarely broadly acuminate with the actual apex obtuse, coriaceous; ovary glabrous ; young branchlets velutinous . . 13. *S. angolensis*

Large, medium-sized or small trees or shrubs with no tendency towards climbing ; tendrils absent :

Lobes of the corolla shorter than the tube ; fruits with hard rind, up to 7 cm. in diameter ; seeds ∞ or 7–12 :

Leaves papery, acuminate ; cymes terminal ; calyx-lobes narrowly triangular to linear ; stamens inserted near the base of the corolla-tube, included ; filaments present ; anthers bearded . . . . . . 2. *S. congolana*

Leaves subcoriaceous, truncate or less often acute ; cymes axillary ; calyx-lobes ± rounded above ; stamens in-

* The fruits of 15, *S. boonei* have not been observed.

serted in the throat of the corolla-tube, just exserted ; anthers sessile, not bearded :
Bark brown, powdery ; leaves 5–15 × 3–8 cm. ; corolla-tube 5 mm. long . . . 10. *S. innocua*
Bark very dark (sometimes almost black), not powdery ; leaves 2–7 × 1–4 cm. ; corolla-tube 3·5 mm. long . . . . 11. *S. dysophylla*
Lobes of the corolla subequal to or longer than the tube ; fruits with thin rind, up to 2 cm. in diameter ; seeds 1(–3) :
Medium-sized to large trees, 6–18 m. tall ; buds more than 3 mm. long ; stamens inserted below the sinuses of the corolla-lobes :
Inflorescence-axes regularly hairy ; corolla cream-coloured ; fruit yellow to orange . . . . . 7. *S. mitis*
Inflorescence-axes only very sparsely hairy if at all ; corolla white ; fruit purple-black . . . 20. *S. stuhlmannii*
Shrubs or small trees, 2–12 m. tall ; buds 2·3–3 mm. long ; stamens inserted in the sinus of the corolla-lobes : *
Lenticels on young branchlets obscure, few and elongated ; seed coffee-bean shaped . . . . . 19. *S. henningsii*
Lenticels on young branchlets distinct, numerous and (after the loss of the epidermis) almost perfectly circular ; seed ellipsoid, subtetragonal or subglobose :
Leaves obovate or broadly elliptic, less than twice as long as broad, strongly coriaceous . . . . 18. *S. xylophylla*
Leaves elliptic to narrowly ovate, not less than twice as long as broad, papery or subcoriaceous :
Leaves narrowly acute or longly acuminate, extreme apex acute 21. *S. usambarensis*
Leaves rounded to subacuminate, extreme apex rounded or truncate . . . . . 14. *S. decussata*

1. **S. myrtoïdes** *Gilg & Busse* in E.J. 32 : 178 & fig. (1902) ; F.T.A. 4 (1) : 531 (1903) ; T.T.C.L. : 274 (1949). Type : Tanganyika, Newala District, Mpatila [Makonde] Plateau, near Nyangao, *Busse* 1108 (B, holo.†, EA, iso. !)

Small much branched densely leafy tree, 1·5–5 m. high ; branchlets slender, twiggy, buff-coloured with persistent petiole bases, ultimate ones, at least pubescent with brown ascending hairs. Leaves subsessile ; lamina very small, subcoriaceous, oblong, elliptic or rarely ovate-orbicular, 0·9–2·5 cm. long, 0·4–1·2 cm. wide, rounded to acute, cuneate or rarely rounded at the base, with the midrib and margin very sparsely hirsute, otherwise

* The flowers of 18, *S. xylophylla* have not been observed.

glabrous, venation indistinct. Cyme terminal, 1–5-flowered. Flowers small, 5-merous, scarcely 2 mm. long. Calyx-lobes suborbicular, ciliate. Corolla-lobes thick, cucullate, oblong-deltoid, subequal to the tube, with dense hairs forming a ring at the base. Stamens included, inserted just above the base of the corolla-tube ; filaments short ; anthers glabrous. Ovary ovoid, glabrous ; style very short or absent, glabrous. Fruit globose, 8–9 mm. in diameter, with a thin skin, red. Seed flattened, disk-like, 7–8 mm. in diameter.

TANGANYIKA. Lindi District : Liho R., *Busse* 2857 ! ; District uncertain, " Gorge Road ", May 1943, *Gillman* 1388 !
DISTR. T8 ; known only from south-eastern Tanganyika and Portuguese East Africa.
HAB. *Brachystegia* woodland ; 500 m.

2. **S. congolana** *Gilg* in E.J. 28 : 120 (1899) ; F.T.A. 4 (1) : 521 (1903) ; E. A. Bruce in K.B. 1955 : 38 (1955). Type : Belgian Congo, near Kasongo, *Dewèvre* 931 (BR, iso. !)

Small tree with pendulous branches, scandent shrub or climber, glabrous except for occasional thin pubescence on the inflorescence and in the angles of the main lateral nerves ; branchlets dark-coloured, not lenticellate, very rarely spiny, sometimes bearing bifurcate tendrils ; ultimate ones paler. Leaves petiolate ; petiole 4–6 mm. long ; lamina membranous, ovate, elliptic or oblong-elliptic, 5–12 cm. long, 2·5–6 cm. wide, rather abruptly acuminate, rounded to cuneate at the base, 3-nerved from 8–12 mm. above the base ; all nerves inconspicuous or slightly impressed above, prominent beneath. Cymes compound, terminal, glabrous or thinly pubescent. Flowers 5-merous. Calyx lobed almost to the base ; lobes narrowly triangular to sublinear, shorter than or rarely equal to the corolla. Corolla greenish ; lobes deltoid, not more than half the length of the tube with dense hairs at the base. Stamens inserted at the base of the corolla-tube ; filaments subequal to the densely bearded anthers. Ovary ovoid ; style short. Fruit globose, about 7 cm. in diameter, with a woody rind. Seeds numerous, compressed, disk-like, about 1·5 cm. diameter, immersed in pulp.

UGANDA. Mengo District : Kampala–Entebbe road, Nov. 1931 (fl.), *Lab. staff in Snowden* 2369 !
DISTR. U4 ; French Guinea, Sierra Leone, Ivory Coast and Belgian Congo.
HAB. Lowland rain-forest ; 1300 m.

VARIATION. The West African material has rather longer and narrower calyx-lobes and a more cuneate leaf-base.

NOTE. The above conception of *S. congolana* includes the West African *S. djalonis* A. Chev. and *S. lecomptei* A. Chev., as well as *S. viridiflora* De Wild. from the Belgian Congo.

3. **S. cocculoïdes** *Baker* in K.B. 1895 : 98 (1895) ; F.T.A. 4 (1) : 533 (1903) ; T.T.C.L. : 275 (1949) ; Lejeunia 13 : 114 (1949) ; B.S.B.B. 85 : 20, fig. 6A (1952) ; E. A. Bruce in K.B. 1955 : 38 (1955). Type : Angola, Huilla, *Welwitsch* 4779 (K, holo. !, BM, iso. !)

Shrub or small tree 2·5–6·5 m. high or, very rarely, climbing (not spreading), with longitudinally ridged, thick, corky branches ; young branchlets often reddish or blackish purple, densely spreading-pubescent or rarely glabrous, usually longitudinally fissured ; spines normally present in pairs in the leaf-axils, ± stout, curved at the tip. Leaves shortly petiolate ; lamina coriaceous, oblong-elliptic to broadly ovate, usually broadest below the middle, 1·8–8 cm. long, 1·4–6 cm. wide (smaller leaves occurring on burnt or grazed shoots), rounded, acute or rarely emarginate at the apex and sometimes apiculate, rounded, subcordate or rarely cuneate at the base,

3–7-nerved at or just above the base (Fig. 3/3), matt or shining above ; venation impressed above, prominent and conspicuous beneath, usually fairly softly pubescent, at least on the nerves ; tertiary nerves visible beneath. Cymes terminal, ± dense, usually pubescent (in our area). Calyx-lobes narrowly triangular to linear-lanceolate, about 3 mm. long, pubescent on the back, slightly shorter than or rarely equal to the corolla. Corolla greenish-white ; lobes deltoid, about half the length of the corolla-tube, with dense hairs at the base. Stamens inserted near the base of the corolla-tube ; anthers densely bearded, longer than the filaments. Ovary ovoid, inconspicuously bilocular ; ovules numerous. Fruit globose, 1·6–7 cm. in diameter, with a smooth woody rind, in the fresh state dark green with paler mottlings. Seeds up to 2 cm. across, compressed but neither so flattened nor so numerous as in *S. spinosa.* Fig. 3/1–6, p. 18.

TANGANYIKA. Dodoma District : Manyoni, Kazikazi, 20 Nov. 1932 (young fr.), *B. D. Burtt*, 4593 ! ; Iringa District : Tanangozi, *Gordon Brown* H54/33/11 ! ; Mbeya District : Mbozi, 21 Nov. 1932 (fl.), *Jessel* 69 !

DISTR. T1, 4–8 ; Belgian Congo, Nyasaland, Northern and Southern Rhodesia, ? Portuguese East Africa, Angola and the Transvaal.

HAB. Deciduous woodland, often on sand ; 400–2000 m.

SYN. *S. goetzei* Gilg in E.J. 28 : 123 (1899) ; F.T.A. 4 (1) : 534 (1903) ; T.T.C.L. : 277 (1949). Type : Tanganyika, Iringa/Ulanga District, Utschungwe [Uzungwa] Mts., *Goetze* 643 (B, holo. !)

S. suberifera Gilg & Busse in E.J. 36 : 107 (1905) ; T.T.C.L. : 275 (1949). Types : Tanganyika, near Lindi, Mayanga, *Busse* 2524 (BM, isosyn. !) & 2524a (B, syn.†, HBG, isosyn. !)

[*S. schumanniana* sensu Brenan, T.T.C.L. : 275 (1949), *non* Gilg]

NOTE. The above conception of *S. cocculoïdes* includes *S. dekindtiana* Gilg, *S. paralleloneura* Gilg & Busse, *S. thomsiana* Gilg & Busse from Angola and *S. suberosa* De Wild. from the Belgian Congo.

4. **S. spinosa** *Lam.*, Illustr. 2 : 38 (1797) ; F.T.A. 4 (1) : 536 (1903), excl. spec. Welw. ex Loanda ; F.W.T.A. 2 : 22, fig. 186 (1931) excl. syn. *S. djalonis* A. Chev. & spec. Chev. ; A. Chev. in Rév. Bot. Appl. 27 : 355 (1947) ; T.T.C.L. : 277 (1949) ; Lejeunia 13 : 109 (1949) ; E. A. Bruce in K.B. 1955 : 40 (1955). Type : Madagascar, unlocalized, unknown collector (P–L, holo., K, photo. !)

Shrub or small tree with spreading habit, up to 6·5 m. high ; bark sometimes reticulate and irregularly corky but never thick ; young branchlets varying from pale to dark mottled or concolorous, glabrous to shortly pubescent, sometimes longitudinally ridged and often powdery corky ; spines slender to stout, straight or curved, sometimes absent. Leaves shortly petiolate ; lamina membranous to coriaceous, ovate, obovate or elliptic, 3–8 cm. long, 1·3–7 cm. wide, rounded, subacute or more rarely acute or emarginate and often apiculate at the apex, cuneate or more rarely rounded at the base, matt or shiny, 3–7-nerved at or just above the base, with hair-pockets sometimes visible in the angles, indumentum of lamina varying on both surfaces from pubescent to glabrous. Cymes terminal ; peduncle and pedicels spreading-pubescent. Calyx-lobes narrowly deltoid to linear, 3·5–5·5 mm. long, shorter or longer than the corolla, acuminate, usually glabrous except for the base, very rarely pubescent up to the lobes. Corolla greenish-white ; lobes deltoid, subacute, about half the length of the corolla-tube, with dense hairs forming a ring at the base. Anthers densely bearded at the base. Ovary broadly ovoid, unilocular ; ovules numerous. Fruit globose, 5–12 cm. in diameter ; rind woody, yellow ; seeds numerous, flat and disk-like, 1·5–2 cm. long, embedded in fleshy yellow edible pulp.

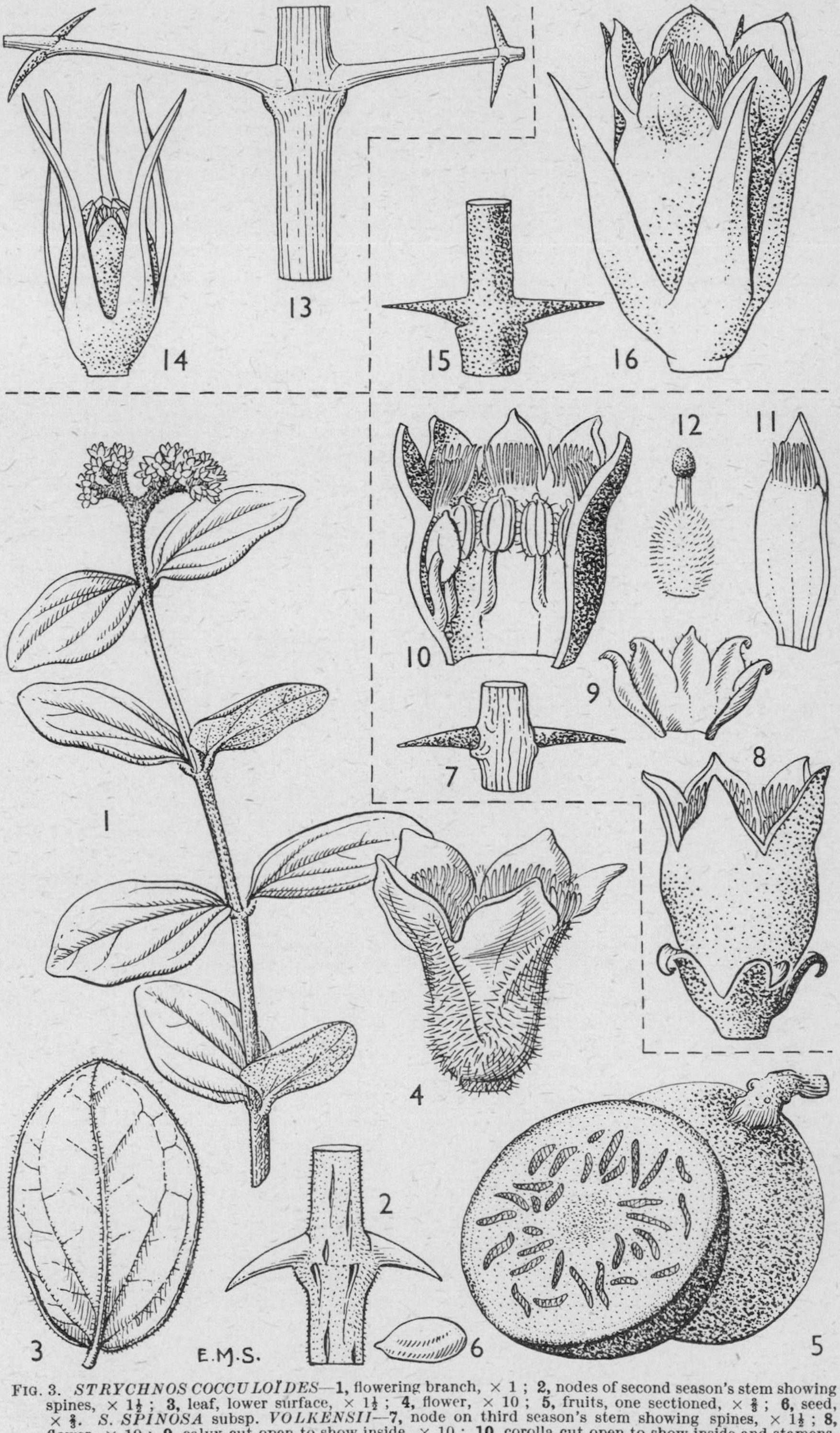

FIG. 3. *STRYCHNOS COCCULOÏDES*—**1,** flowering branch, × 1 ; **2,** nodes of second season's stem showing spines, × $1\frac{1}{2}$; **3,** leaf, lower surface, × $1\frac{1}{2}$; **4,** flower, × 10 ; **5,** fruits, one sectioned, × $\frac{2}{3}$; **6,** seed, × $\frac{2}{3}$. *S. SPINOSA* subsp. *VOLKENSII*—**7,** node on third season's stem showing spines, × $1\frac{1}{2}$; **8,** flower, × 10 ; **9,** calyx cut open to show inside, × 10 ; **10,** corolla cut open to show inside and stamens, 10 ; **11,** corolla-lobe and part of tube, from within, × 10 ; **12,** pistil, × 10. *S. SPINOSA* subsp. *SPINOSA*—**13,** node of third season's stem showing spiny lateral branches, × $1\frac{1}{2}$; **14,** flower, × 10 ; *S. SPINOSA* subsp. *LOKUA*—**15,** node of second season's stem showing spines, × $1\frac{1}{2}$; **16,** flower, × 10. 1, 3, 4, from *C. H. N. Jackson* 119 ; 2, from *B. D. Burtt* 4593 ; 5, 6, from *White* 1874 ; 7–12, from *Bally* 5976 ; 13, 14, from *Greenway* 5213 ; 15, from *Eggeling* 1509 ; 16, from *Koritschoner* 1718.

KEY TO SUBSPECIES

Young branchlets with a ± varnished surface, usually dark or mottled ; leaves usually ± membranous, with inconspicuous tertiary nerves :
- Calyx-lobes equal to or longer than the corolla ; inflorescence fairly dense ; spines slender, ± straight ; leaves glabrous except for angles of nerves beneath . . . subsp. **spinosa**
- Calyx-lobes much shorter than the corolla ; inflorescence rather lax ; spines straight or curved ; leaves glabrous or more rarely pubescent . . . subsp. **volkensii**

Young branchlets neither varnished nor mottled, usually pale and powdery corky ; spines ± stout, curved at the apex ; leaves coriaceous ; tertiary nerves conspicuous ; leaves usually glabrous, but occasionally pubescent . . . . subsp. **lokua**

subsp. **spinosa** ; E. A. Bruce in K.B. 1955 : 40 (1955)

Young branchlets glabrous, dark or mottled, ± varnished ; branchlets white or pale buff, smooth, hard, not powdery corky ; spines slender, straight or nearly so, not recurved at the tip. Leaves usually membranous, glabrous except for the hair-pockets in the angles ; tertiary nerves inconspicuous beneath. Cymes fairly dense. Calyx-lobes linear, longer than the corolla. Fig. 3/13.

TANGANYIKA. Near Dar es Salaam, *Raymond* 23 ! ; Rufiji District : Mafia Island, Mito Miwele, 9 Aug. 1937 (fl. & fr.), *Greenway* 5026 ! ; Lindi District : 1·5 km. N. of Mchinga, 7 Dec. 1955 (young fl.), *Milne-Redhead & Taylor* 7583 !

DISTR. **T**6, 8 ; coastal parts of Portuguese East Africa, also in Swaziland, the Transvaal, Natal, eastern and southern Cape Province in South Africa as well as Madagascar, the Mascarene Islands and the Seychelles.

HAB. Coastal evergreen bushland, often persisting in coconut plantations and induced grassland ; 6–12 m.

NOTE. This plant has been called *S. carvalhoi* Gilg (in Portuguese East Africa) and *S. madagascariensis* Poir.

subsp. **volkensii** (*Gilg*) *E. A. Bruce* in K.B. 1955 : 40 (1955). Types : Tanganyika, Tanga, *Holst* 2095 (B, syn.†, K, isosyn. !) & *Volkens* 103 (B, syn.†, BM, isosyn. !)

Young branchlets as in subsp. *spinosa;* leaves glabrous or pubescent on the nerves below ; spines, when present ± slender, straight or curved. Cymes fairly lax. Calyx-lobes much shorter than the corolla. Fig. 3/7–12.

KENYA. Kwale, *R. M. Graham* 1770 ! ; Lamu District : Kiunga, 15 Dec. 1946 (fl.), *Mrs. J. Adamson* 285 *in Bally* 5976 ! & Witu, *F. Thomas* 198 !

TANGANYIKA. Lushoto District : Korogwe–Kwata road near Kerengi, 29 June 1953 (fr.), *Drummond & Hemsley* 3071 ! ; Tanga District : 8 km. SE. of Ngomeni, 2 Aug. 1953 (fr.), *Drummond & Hemsley* 3612 ! & Sawa, 23 Apr. 1956 (fl.), *Faulkner* 1856.

ZANZIBAR. Zanzibar Is., Marahubi, 25 Sept. 1930 (fl.), *Vaughan* 1574 !

DISTR. **K**7 ; **T**3, 8 ; **Z** ; not known elsewhere.

HAB. Coastal evergreen bushland, often persisting in coconut plantations and induced grassland ; 0–1000 m.

SYN. *S. volkensii* Gilg in P.O.A. C : 311 (1895) ; F.T.A. 4 (1) : 536 (1903) ; T.S.K. : 126 (1936) ; T.T.C.L. : 278 (1949) ; Lejeunia 13 : 111 (1949)

?*S. sansibarensis* Gilg in E.J. 28 : 124 (1899) ; F.T.A. 4 (1) : 535 (1903) ; T.T.C.L. : 277 (1949). Type : Zanzibar, unlocalized, *Stuhlmann* 161 (B, holo.†)

S. megalocarpa Gilg & Busse in E.J. 32 : 180 (1902) ; F.T.A. 4 (1) : 526 (1903) ; T.T.C.L. : 278 (1949). Type : Tanganyika, Handeni District, Kwa Sseulanga [Kwa-Zuranga], *Busse* 323 (B, holo.†)

S. omphalocarpa Gilg & Busse in E.J. 32 : 181 (1902) ; F.T.A. 4 (1) : 525 (1903) ; T.T.C.L. : 278 (1949). Type : Tanganyika, Handeni District, Kwa Mdoë [near Handeni], *Busse* 322 (B, holo.†)

S. radiosperma Gilg & Busse in E.J. 36 : 108 & t. 2c (1905) ; T.T.C.L. 275 (1949). Type : Tanganyika, Kilwa District, Matumbi Mts., near Mirungamo, *Busse* 3061 (B, holo.†, BM, EA, HBG, iso. !)

S. cuneifolia Gilg & Busse in E.J. 36 : 109 (1905) ; T.T.C.L. : 273 (1949). Type : Tanganyika, Lindi District, Lake Lutamba, *Busse* 2519 (B, holo.†, EA, iso. !)
[*S. spinosa* sensu T.S.K. : 125 (1936), *non* Lam.]

subsp. **lokua** (*A. Rich.*) *E. A. Bruce* in K.B. 1955 : 42 (1955). Type : Ethiopia, Tigré, Takkaze River, *Quartin Dillon* (P, holo. !)

Young branchlets pale, matt, not varnished ; older branchlets with powdery bark ; spines, when present, fairly stout, curved at the apex. Leaves usually coriaceous, with conspicuous tertiary nerves beneath, glabrous or more rarely pubescent. Calyx-lobes variable, intermediate in length between subspp. *spinosa* & *volkensii*. Fig. 3/15, 16, p. 18.

UGANDA. West Nile District : Mt. Ite, Feb. 1934 (fl. and fr.), *Eggeling* 1509 *in F.D.* 1439 ! ; Acholi District : Agoro, Mar. 1935 (fl.), *Eggeling* 1715 *in F.D.* 1604 ! ; Teso District : Serere, Aug. 1932 (sterile), *Chandler* 846 !
KENYA. West Suk District : Kapenguria, *Hale* 36 *in Bally* 4051 ! ; Machakos District : about 11 km. S. of Kangundo, 18 Jan. 1954 (young fr.), *Hemming* 238 ! ; Kitui Boma, 18 Jan. 1942 (fl.), *Bally* 1706 !
TANGANYIKA. Mwanza District : Mabale, Mbarika, 17 May 1953 (fr.), *Tanner* 1428 ! ; Tabora District : Chigunga, 20 Oct. 1949 (fl.), *Leonard* 21 ! ; Iringa District : Sao Hill, Oct. 1935 (fl.), *Hornby* 664 !
DISTR. **U**1, 3 ; **K**2, 4 ; **T**1, 3–8 ; from Gambia eastwards to the Sudan then southwards through eastern Africa to Angola, the Transvaal and Cape Province, South Africa.
HAB. Deciduous woodland and upland scattered tree grassland ; 400–2200 m.

SYN. *S. lokua* A. Rich., Tent. Fl. Abyss. 2 : 53 (1851) ; Lejeunia 13 : 112 (1949)
S. tonga Gilg in E.J. 17 : 575 (1893) ; F.T.A. 4 (1) : 527 (1903) ; T.T.C.L. : 275 (1949). Types : Portuguese East Africa, Quilimane, *Stuhlmann* 1039 (B, syn.†) ; Tanganyika, Pangani, *Stuhlmann* (B, syn.†)
S. euryphylla Gilg & Busse in E.J. 32 : 179 (1902) & in E.J. 36 : 102, 108, t. 2a (1905) ; F.T.A. 4 (1) : 526 (1903) ; T.T.C.L. 274 (1949). Types : Tanganyika, Uluguru Mts., Kiroka, *Busse* 146 (B, syn.†) & Usgara Mts., Kilosa, *Busse* 174 (B, syn.†) & Songea District, near Kwa Lituno [near R. Rovuma 27 km. W. of Songea], *Busse* 1263 (B, syn.†, EA, isosyn. !)
S. cardiophylla Gilg & Busse in E.J. 36 : 110 (1905) ; T.T.C.L. : 273 (1949). Type : Tanganyika, Kilwa District, Singino Hill [just S. of Kilwa Kavinje], *Busse* 3011 (B, holo.†, EA, iso. !)
S. harmsii Gilg & Busse in E.J. 36 : 109 (1905) ; T.T.C.L. : 274 (1949). Types : Tanganyika, Lindi District, E. Muera [Rondo] Plateau, *Busse* 2303 (B, syn.†) & Rondo Plateau, *Busse* 2560 & 2596 (B, syn.†, EA, isosyn. !)
?*S. mueghe* Chiov. in Racc. Bot. Miss. Consol. Kenya 83 (1935). Types : Mt. Kenya, " Ottune Steppe," *Balbo* 686 (TOM, syn.) & " Saraka Steppe," *Balbo* 687 (TOM, syn.)
[*S. spinosa* sensu Dale, I.T.U., ed. 2 : 169 (1952), *non* Lam.]

NOTE. Ten other ' species ' from outside our area fall under this subspecies (see K.B. 1955 : 42 (1955)).
The placing of the synonyms under the subspecies may be uncertain for various reasons, for example, type material may be absent or inadequate (see K.B. 1955 : 41 & 44 (1955)). Good correlated material of flowers and fruit is necessary before a further analysis of this species can be undertaken.

VARIATION. There is very great variation within this species and within the subspecies themselves, particularly in leaf-shape, leaf-size and pubescence. At present the confusing variability of the subspecies is presumed to be due to intermittent hybridization rather than to the presence of stable infrasubspecific entities.

5. **S. aculeata** *Solered.* in E. & P. Pf. 4 (2) : 40 (1892) & in E.J. 17 : 556 (1893) ; F.T.A. 4 (1) : 520 (1903) ; F.W.T.A. 2 : 22 (1931) ; B.S.B.B. 85 : 21 (1952). Type : Fernando Po, *Mann* 175 (B, holo.†, K, iso. !)

Lofty woody climber, glabrous in all its parts. Stems 5–30 m. long and up to 20 cm. in diameter at the base ; young branchlets subquadrangular, pale brown and shining or dull purple-brown, usually bearing flattened recurved prickles along the edges ; older branchlets longitudinally fissured, with scattered prickles except at the base ; tendrils usually present, pedunculate, in 1–3 pairs. Leaves petiolate ; petiole 5–10 mm. long ; lamina coriaceous, oblong-elliptic to broadly elliptic, 8–16 cm. long, 4·5–7 cm. wide, acute or rounded and then abruptly acuminate, cuneate or rounded at the base, 3-nerved from just above the base ; nerves impressed above, prominent below ; subsidiary lateral nerves arising at wide angles. Cymes axillary, many-flowered ; peduncle 1–3 cm. long. Flowers 5-merous.* Calyx-lobes suborbicular, wrinkled, ciliate. Corolla greenish ; lobes thick, deltoid, about half the length of the tube, with dense hairs forming a ring at the base. Anthers thinly bearded, subsessile, inserted about the middle of the corolla-tube. Ovary ovoid ; stigma sessile, oblong, bilobed. Fruits globose, 9–12 cm. in diameter, with a woody rind, yellow. Seeds numerous, large, compressed, about 3 cm. in diameter, poisonous.

Uganda. Bunyoro District : Budongo Forest, Oct. 1938 (fr.), *Eggeling* 3817 !
Distr. U2 ; Sierra Leone, eastwards through the Cameroons, Gabon, Belgian & Portuguese Congo to Uganda.
Hab. Lowland rain-forest ; 1200 m.

Syn. *S. mortehani* De Wild. in B.J.B.B. 5 : 50 (1915). Type : Belgian Congo, Dundusana, *Mortehan* 485 (BR, holo. !)

6. **S. matopensis** *S. Moore* in J.B. 43 : 48 (1905) ; Lejeunia 11 : 77 (1947) ; T.T.C.L. : 274 (1949) ; E. A. Bruce in K.B. 1956 : 156 (1956). Type : Southern Rhodesia, Matopo Hills, *Eyles* 1182 (BM, holo. !)

Woody climber or usually scandent shrub about 4 m. high, forming a dense tangle ; branches rather densely grey to tawny pubescent, not obviously lenticellate, with rather close, spreading, opposite and decussate branchlets ; tendrils usually present, simple, axillary, pubescent. Leaves subsessile or shortly petiolate ; lamina subcoriaceous, usually broadly ovate, more rarely ovate-lanceolate or elliptic, 1–5 cm. long, 1–2·7 cm. wide, acute or subacuminate, rarely rounded and mucronulate, rounded to cuneate at the base, 3–5-nerved from the base and with a marginal nerve, glabrous or with a few hairs along the midrib and the margins towards the base, often drying dark-coloured, but paler beneath ; nerves impressed above, prominent beneath ; tertiary venation not conspicuous. Cymes terminal and axillary, many-flowered, congested, pubescent. Flowers 5-merous. Calyx-lobes broadly ovate, rounded and shortly acuminate, ciliate and pubescent on the back. Corolla pale yellow ; lobes deltoid, rather thick ; tube narrowly campanulate, longer than the lobes, thinly pubescent outside, with dense hairs forming a ring at the throat ; stamens inserted in the upper part of the corolla-tube below the sinuses ; anthers large, subtriangular, bearded at the base, not exserted. Style glabrous, usually longer than the glabrous subglobose ovary. Fruits ellipsoid, about 1 cm. diameter, yellow to orange. Seeds 1–2, elliptic, compressed, 1 cm. long, 0·6 cm. wide.

Tanganyika. Ufipa District : Kasanga, 17 Mar. 1932, *H. E. Hornby* 1018 ! ; Kahama District : Mkweni, 21 Nov. 1938 (fl.), *Glover* 159 ! ; Singida District : Rift Valley, near Mau Hills, Sept. 1935 (sterile), *B. D. Burtt* 5227 !
Distr. T1, 4, 5 ; Northern & Southern Rhodesia.
Hab. Semi-evergreen bushland on rocky hills and on termite-hills ; 1200–160 m.

7. **S. mitis** *S. Moore* in J.L.S. 40 : 146 (1911) ; I.T.U., ed. 2 : 167 (1952) ; B.S.B.B. 85 : 24 (1952). Type : Southern Rhodesia, Chirinda Forest, *Swynnerton* 17a (BM, holo. !, K, iso. !)

Tall branched, evergreen tree with rounded crown, 10–35 m. high (rarely shorter), with neither spines nor tendrils ; bark smooth, grey ; branchlets pale grey or buff, lenticellate, ascending, glabrous or more rarely pubescent. Leaves shortly petiolate or subsessile, subcoriaceous, broadly to narrowly

* The flowers were described in F.T.A. as 4-merous with corolla-lobes longer than the tube, but this is not found in the original description nor is it confirmed by dissection.

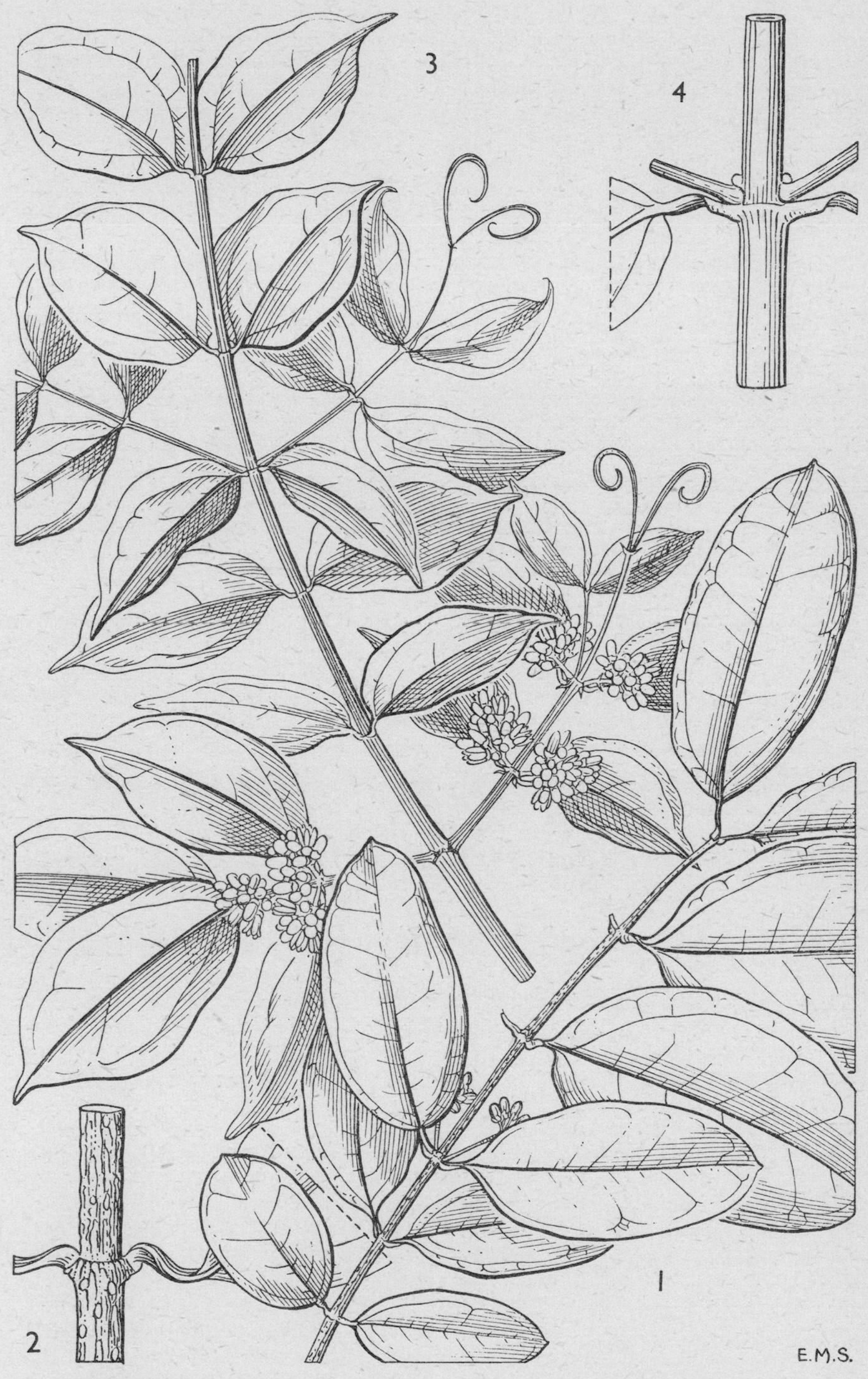

Fig. 4. *STRYCHNOS LUCENS*—**1,** flowering branch, $\times \frac{2}{3}$; **2,** node of stem showing lenticels, $\times 2$; *S. SCHEFFLERI*—**3,** flowering branch, $\times \frac{2}{3}$; **4,** node of stem, $\times 2$. 1, 2, from *Lindeman* 733 ; 3, 4, from *Drummond & Hemsley* 3354.

elliptic, more rarely ovate to ovate-lanceolate, 4–11 cm. long, 1·7–5 cm. wide, generally acuminate, more rarely acute or obtuse, cuneate at the base, glabrous above, glabrous or with angle hairs or pubescent beneath, 3-nerved from about 1 cm. above base, with a fainter submarginal nerve from the base ; tertiary nerves spreading, not very prominent. Cymes axillary and terminal, compound, dichasial, fairly dense, 1–2·5 cm. long, pubescent with short spreading hairs ; peduncle usually short, rarely up to 4 cm. long. Flowers 4–5-merous, subsessile, about 3 mm. long. Calyx-lobes broadly ovate to suborbicular, about 1·7 mm. long, pubescent on the back, ciliate. Corolla cream, yellow or greenish ; tube shortly campanulate, glabrous ; lobes ovate-deltoid, subequal to the tube or slightly longer with dense hairs at the base. Stamens inserted about the middle of the corolla-tube on short filaments ; anthers just exserted. Ovary ovoid, slightly longer than the style. Fruit subglobose, 1·2–2 cm. long, with a thin yellow to orange rind ; calyx and bracteoles persistent. Seeds 1(–2), subellipsoid, usually flattened on one side, 1–1·2 cm. long ; dried-up mesocarp often adhering to seed-coat as a rugose or foveolate covering ; hilum often conspicuous in centre of ellipsoid surface.

UGANDA. Karamoja District : Napak Camp, Feb. 1938 (fr.), *Sangster* 425, 426 ! ; Toro District : Bwamba, Kidongo, Aug. 1937 (fr.), *Eggeling* 3378 ! ; Mengo District : Kyagwe, Mabira Forest, Mar. 1908 (fr.), *Usher* 14 !

KENYA. Baringo District : Kamasia Hills, Bartolimo Forest, Apr. 1941, *Wimbush* 1220 ! ; Nairobi Arboretum, June 1940 (fl.), *Bally* 893 !

TANGANYIKA. Bukoba District : Minziro Forest Reserve, Kabobwa, Sept. 1950, *Watkins* 514 *in F.D.* 3253 ! ; Arusha District : Tengeru, Jan. 1953 (fl.), *Eggeling* 6482 ! ; Lushoto District : W. Usambara Mts., on Lushoto–Mombo road, 1 km. NE. of Vuga turn-off, 14 June 1953 (fl.), *Drummond & Hemsley* 2919 !

DISTR. **U**1, 2, 4 ; **K**3, 4, 6 ; **T**1–3, 6 ; also in south-eastern Sudan and Southern Rhodesia.

HAB. Upland and lowland rain-forest and riverine forest ; 0–2300 m.

SYN. *Strychnos* sp. in T.S.K. : 126 (1936)

NOTE. From the same type locality Spencer Moore also described *S. mellodora*, which scarcely differs from our species ; they may well come to be considered conspecific.

8. **S. lucens** *Baker* in K.B. 1895 : 97 (1895) ; F.T.A. 4 (1) : 524 (1903) ; Bruce & Lewis in K.B. 1956 : 269 (1956). Type : Angola, Loanda District, *Welwitsch* 6015 (BM, holo. !, K, iso. !)

Woody climber or semi-scandent shrub or small tree with sarmentose branches, 5–20 m. high ; branchlets brown to greyish-straw-colour, glabrous or rarely with short ± spreading stiff ginger-coloured hairs on the ultimate branchlets (*Tanner* 1557) ; tendrils sometimes present, bifurcate ; lenticels usually conspicuous, generally elongate, more rarely circular ; petiole-bases persistent, ± prominent. Leaves shortly petiolate ; lamina thickly coriaceous, oblong-elliptic to broadly elliptic, very variable in size, 2–11 cm. long, 1–7 cm. wide, acute, shortly acuminate or more rarely rounded or subtruncate at the apex, rounded or broadly cuneate at the base, usually shining above, shining or ± dull beneath, strongly 3-nerved from or just above the base ; faint marginal nerve often present ; tertiary nerves arising at a wide angle to the midrib ; venation conspicuously prominent and reticulate on both surfaces, glabrous or rarely with a few hairs near the base beneath. Cymes axillary, up to 2·5 cm. long, shortly pedunculate, 3–9-flowered, pubescent with brown hairs, Flowers 4–5-merous, about 6 mm. long. Calyx-lobes elliptic to broadly elliptic, rounded, ciliate, definitely shorter than corolla-tube. Corolla greenish-white ; tube subcylindrical ; lobes oblong, acute, thickened and cucullate at the apex, shorter than the tube ; corolla glabrous except for dense hairs forming a ring in the throat. Stamens sessile, inserted at mouth of corolla ; anthers glabrous, just exserted. Style

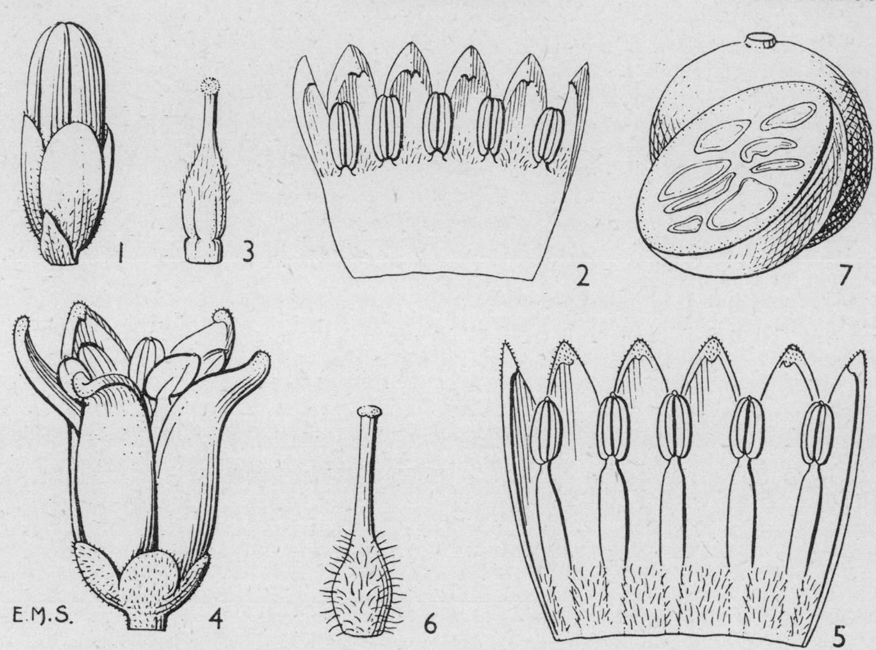

FIG. 5. *STRYCHNOS LUCENS*—**1**, flower-bud, × 6; **2**, corolla opened out to show stamens, × 6; **3**, pistil, × 6; *S. SCHEFFLERI*—**4**, flower, × 6; **5**, corolla opened out to show stamens, × 6; **6**, pistil, × 6; **7**, fruit × 1. 1–3, from *Lindeman* 733; 4–7 from *Drummond & Hemsley* 3354.

gradually tapered into the ovary, densely pilose below, ± subequal to the corolla-tube. Fruit 1·5–5 cm. in diameter, with firm rind, orange. Seeds up to 30, thick, ovoid, about 1–1·3 cm. long. Fig. 4/1, 2, p. 22; Fig. 5/1–3.

TANGANYIKA. Mwanza District: Geita, Ibondo, 4 July 1953 (fr.), *Tanner* 1557!; Tabora District: Ugalla River, 31 Oct. 1938 (fr.), *Lindeman* 733!; Mpwapwa District: Kikombo, 19 Aug. 1933 (sterile), *B. D. Burtt* 4794!
DISTR. **T**1, 4, 5, 7; Northern Rhodesia & Angola.
HAB. Riverine forest & semi-evergreen bushland on rocky hills; 1050–1700 m.

SYN. *S. milneredheadii* Duvign. & Staq. in B.S.B.B. 84: 69 (1951). Type: Northern Rhodesia, Mwinilunga District, R. Matonchi, *Milne-Redhead* 2947 (BR, holo.!, K, iso.!)

VARIATION. This species varies greatly in size of leaf and fruit, the type from Angola having smaller leaves and fruits than the majority of specimens from Tanganyika & Northern Rhodesia.

9. **S. pungens** *Solered.* in E. & P. Pf. 4 (2): 40 (1892); E.J. 17: 554 (1893); P.O.A. C: 310 (1895); F.T.A. 4 (1): 530 (1903); Duvign. in B.S.B.B. 85: 25 (1952); Bruce & Lewis in K.B. 1956: 268 (1956). Types: Angola, Huilla, *Welwitsch* 4778 (B, syn.†, BM, K, isosyn.!) & Tanganyika, Dodoma District, Saranda, *Fischer* 374 (B, syn.†, EA, K, isosyn.!)

Shrub or slender much-branched small tree, 3–8 m. high, with a ± dense stiff growth, glabrous except for the inflorescence; branches unarmed, with rough grey or brown bark, often longitudinally split or corky; young branchlets usually pale brown, conspicuously lenticellate. Leaves subsessile, coriaceous, rigid, narrowly oblong-lanceolate, elliptic or obovate, 4–8 cm. long, 1–4 cm. wide, acute or subacute and conspicuously pungent, cuneate or rounded at the base, shining above; nerves 3 (or 5) from or just above the base; tertiary venation conspicuous. Cymes usually ± dense and shorter than the leaves, rarely lax. Calyx-lobes ovate, acute, ciliate, about

2·5 mm. long. Corolla greenish-cream-coloured ; tube 3–4 mm. long, with dense hairs forming a ring at the throat ; lobes oblong, 2–3 mm. long. Anthers sessile, inserted at mouth of corolla, exserted. Ovary subglobose ; style long ; ovules numerous. Fruit globose, 5–9 cm. in diameter ; rind woody, yellow. Seeds numerous, rather thick, tetrahedral, about 2 cm. long, embedded in sweet-tasting pulp.

TANGANYIKA. Tabora District : Kaliuwa, 17 Oct. 1949 (fl.), *Ramazani* 26 ! ; Singida District : road to Muwera, 7 Mar. 1928 (fr.), *B. D. Burtt* 1385 ! ; Songea District : 2·5 km. E. of Johannesbruck, 19 Apr. 1956 (fr.), *Milne-Redhead & Taylor* 9763 !
DISTR. T4, 5, 8 ; Belgian Congo, Nyasaland, Northern & Southern Rhodesia, Angola, the Transvaal & South West Africa.
HAB. *Brachystegia* woodland ; 900–1350 m.

NOTE. Included within this species are *S. occidentalis* Solered. and *S. henriquesiana* Baker from Angola, and *S. sapini* De Wild. from the Belgian Congo.

10. **S. innocua** *Del.*, Cent. Pl. Afr. : 53 (1826) ; F.T.A. 4 (1) : 532 (1903) ; Bullock & E. A. Bruce in K.B. 1938 : 54 (1938) ; T.T.C.L. : 274 (1949) ; I.T.U., ed. 2 : 167, photo. 29, fig. 35 (1952) ; Fl. Pl. Sudan 2 : 381 (1952) ; Bruce & Lewis in K.B. 1956 : 270 (1956). Type : from Ethiopia, observed or collected by *Calliaud*.

Shrub or small tree, 2–12 m. high, with yellowish-white or buff-coloured farinose branchlets, usually without lenticels. Leaves subsessile or shortly petiolate, obovate, elliptic or oblong-elliptic, 4·5–15 cm. long, 2–8 cm. wide, coriaceous, rounded-emarginate or rarely subacute, widely to very narrowly cuneate or rarely rounded at the base, glabrous to pubescent, matt or shining above, matt beneath, 3–7-nerved from near the base ; tertiary venation prominent at least beneath. Cymes axillary, pedunculate, usually pubescent. Calyx-lobes about 3 mm. long, pubescent and ciliate. Corolla greenish, white or yellowish ; tube cylindric, about 5 mm. long, hirsute at the throat ; lobes oblong, thick, cucullate, acute, about 3·5 mm. long. Anthers sessile in the corolla-throat. Ovary elongate-ovoid ; style hirsute below. Fruit globose, 7 cm. or more in diameter ; rind hard, orange or orange-yellow. Seeds tetrahedral, 1·5–1·8 cm. across, thicker and less numerous than those of *S. spinosa*.

KEY TO INTRASPECIFIC VARIANTS

Upper surface of leaves not shining ; tertiary venation prominent (subsp. **innocua**) :
  Leaves glabrous beneath . . . . var. **innocua**
  Leaves pubescent beneath, at least on the midrib at the base . . . . var. **pubescens**
Upper surface of leaves ± shining ; tertiary venation inconspicuous (subsp. **burtonii**) :
  Leaves pubescent beneath, at least on the midrib at the base . . . . var. **burtonii**
  Leaves glabrous beneath . . . . var. **glabra**

subsp. **innocua**; Bruce & Lewis in K.B. 1956 : 271 (1956)

Plants bearing lateral shoots of orthodox annual increase in length. Leaves cuneate or rounded at the base, rarely narrowly cuneate (when young) ; both surfaces matt, with a conspicuous prominent reticulate venation ; one to three pairs of lateral nerves ± prominent and clearly visible on both surfaces.

var. **innocua** ; Bruce & Lewis in K.B. 1956 : 271 (1956)

Branchlets glabrous. Leaves glabrous.

UGANDA. West Nile District : Amua, *Eggeling* 904 *in F.D.* 1252 ! ; Busoga District : unlocalized, *Snowden* 216 ! ; Mengo District : Singo, near Bukomero, Aug. 1932 (fr.), *Eggeling* 483 *in F.D.* 819 !

TANGANYIKA. Mwanza District : Geita, 6 June 1937 (fl. & fr.), *B. D. Burtt* 6571 ! ; Kigoma District : Usinge, 26 Nov. 1933 (fl.), *Michelmore* 770 !
DISTR. **U**1–4 ; **T**1, 4, 7 ; West Africa eastwards to the Sudan & Ethiopia, then south to Southern Rhodesia.

SYN. *S. unguacha* A. Rich., Voy. Abyss. Bot, Atlas, t. 73 (1847) & Tent. Fl. Abyss. 2: 52 (1851) ; P.O.A. C : 310 (1895) ; F.T.A. 4 (1) : 534 (1903). Type : Ethiopia, Tigré, Tacazze valley, *Schimper* 1817 (K, iso. !)
S. unguacha A. Rich. var. *micrantha* Gilg in E.J. 17 : 563 (1893) ; P.O.A. C : 310 (1895). Type : Tanganyika, Pangani, *Stuhlmann* 76 (B, holo.†)
S. unguacha A. Rich. var. *steudneri* Gilg in E.J. 17 : 563 (1893) ; P.O.A. C : 310 (1895). Type : NW. Ethiopia, *Steudner* 852 (B, syn.†)
S. unguacha A. Rich. var. *polyantha* Gilg in E.J. 30 : 374 (1901). Type : Tanganyika, Mbeya District, near Kananda, *Goetze* 1436 (B, holo.†)

var. **pubescens** *Solered.* in E.J. 17 : 556 (1893) ; Bruce & Lewis in K.B. 1956 : 271 (1956). Type : Nigeria, Nupe, *Barter* 1160 (B, holo.†, K, iso.!)

Branchlets pubescent. Leaves pubescent, at least on the midrib at the base beneath.

UGANDA. Acholi District : Adilang, 11 Apr. 1945 (fl.), *Greenway & Hummel* 7339 ! ; Lango District : near Lira, Ngeta, 18 Jan. 1931 (fl. & fr.), *A. W. Hill* 13 ! ; Teso District : Serere, *Eggeling* 753 *in F.D.* 1150 !
TANGANYIKA. Shinyanga District : 18 June 1931 (sterile), *B. D. Burtt* 3301 ! ; Tabora, *Lindeman* 432 ! ; Singida District : near Muwera, 7 Mar. 1928, *B. D. Burtt* 1387 !
DISTR. **U**1, 3 ; **T**1, 4, 5, 7 ; West Africa eastward to the Sudan, then south to Northern Rhodesia and Angola.

SYN. *S. fischeri* Gilg in E.J. 17 : 565 (1893) ; P.O.A. C : 310 (1895) ; F.T.A. 4 (1) : 535 (1903) ; T.T.C.L. : 274 (1949). Type : Tanganyika, Shinyanga District, Usule, *Fischer* 300 (B, holo.†)
S. triclisioïdes Baker in K.B. 1895 : 98 (1895) ; F.T.A. (4) 1 : 533 (1903) ; F.W.T.A. 2 : 22 (1931). Type : Nigeria, Nupe, *Barter* 1160 (K, holo. !)
S. xerophila Baker in K.B. 1895 : 98 (1895) ; F.T.A. 4 (1) : 534 (1903). Types : Uganda, West Nile District, Madi, *Speke & Grant* (K, syn !) & Sudan, Equatoria Province, Seriba Kurshook Ali [near Wau], *Schweinfurth* 1719 (K, syn. !)

subsp. **burtonii** (*Baker*) *Bruce & Lewis* in K.B. 1956 : 272 (1956). Type : Portuguese East Africa, Manica e Sofala, Chupanga, *Kirk* 368 (K, lecto. !)

Lateral shoots usually very short, with numerous congested leaf-scars. Leaves attenuate or narrowly cuneate below, the lamina being narrower on average than in subsp. *innocua* and sometimes scimitar-shaped ; upper surface ± shining with tertiary venation not conspicuously prominent ; only the midrib and one pair of lateral nerves at all conspicuous, the other pair being narrow, not prominent, and submarginal.

var. **burtonii** ; Bruce & Lewis in K.B. 1956 : 273 (1956)

Branchlets pubescent, especially when young. Lamina of leaves pubescent, at least on the midrib towards the base beneath ; petiole pubescent.

TANGANYIKA. Kondoa District : Kolo, 4 Dec. 1925 (sterile), *B. D. Burtt* 312 ! ; Kilosa, 23 Jan. 1926 (sterile), *B. D. Burtt* 88 ! ; Lindi District : Rondo Plateau, Mtene, Nov. 1953 (fl.), *Eggeling* 6724 !
ZANZIBAR. Zanzibar Is., unlocalized, *Burton*
DISTR. **T**3, 5, 6, 8 ; **Z** ; Nyasaland and Portuguese East Africa.

SYN. *S. burtonii* Baker in K.B. 1895 : 98 (1895) ; F.T.A. 4 (1) : 533 (1903)
S. behrensiana Gilg & Busse in E.J. 32 : 175 (1902) ; E.J. 36 : 100 (1905) ; F.T.A. 4 (1) : 531 (1903). Type : Tanganyika, Pangani District, Mt. Tongwe, *Stuhlmann* 76 (B, syn.†)
S. quaqua Gilg in E.J. 17 : 565 (1893) ; P.O.A. C : 310 (1895) ; F.T.A. 4 (1) : 531 (1903). Type : Portuguese East Africa, Quilimane, *Stuhlmann* 1041 (B, holo.†, HBG, iso. !, K, photo.-iso. !)

var. **glabra** *Bruce & Lewis* in K.B. 1956 : 273 (1956). Type : Tanganyika, Tanga District, Kwamkembe–Pongwe, *Greenway* 4851 (EA, holo. !)

Branchlets glabrous, even those of the first year. Leaves glabrous, even on the midrib towards the base beneath.

KENYA. Kwale, Nov. 1928 (fl.), *R. M. Graham* 237 !
TANGANYIKA. Tanga District : 6·5 km. N. of Amboni, Dec. 1935 (sterile), *B. D. Burtt* 5366 ! & 8 km. SE. of Ngomeni, Aug. 1953 (fr.), *Drummond & Hemsley* 3611 !

ZANZIBAR. Zanzibar Is., Masingini Ridge, Feb. 1929 (sterile), *Greenway* 1289 !
DISTR. **K7** ; **T3**, 6 ; **Z** ; Southern Rhodesia & Portuguese East Africa.

SYN. [*S. burtonii* sensu Baker in K.B. 1895 : 98 (1895), pro parte, quoad *Kirk* 96]
S. melonicarpa Gilg & Busse in E.J. 36 : 101 (1905) ; T.T.C.L. : 278 (1949). Type : Tanganyika, Pangani District, near Mnyuzi, *Busse* 2266 (B, holo.†, EA, iso. !)
S. stenoneura Gilg & Busse in E.J. 36 : 103 (1905) ; T.T.C.L. : 278 (1949). Type : Tanganyika, Lindi District, Mayanga, *Busse* 2537 (B, lecto. †, EA, HBG, isolecto. !)
?*S. leiocarpa* Gilg & Busse in E.J. 36 : 103 (1905) ; T.T.C.L. : 278 (1949). Types : Tanganyika, Lindi District, Mtange, *Busse* 2458 (B, syn.†, EA, isosyn.!) & Kilwa District, near Kipunga, *Busse* 2938 (B, syn.†, EA, isosyn. !)
S. polyphylla Gilg & Busse in E.J. 36 : 104 (1905) ; T.T.C.L. 278 (1949). Types : Tanganyika, Kilwa District, Matumbi Mts., *Busse* 3058 (B, syn.†, BM, EA, & HBG, isosyn. !) & near Kwakikumba, *Busse* 3063 (B, syn.†, EA, isosyn. !)

HAB. (for species as a whole) ; deciduous woodland ; up to 1400 m.

NOTE. The queried synonyms are less certainly placed than the rest because the characters of the twigs of their type specimens show affinity with *S. dysophylla* subsp. *engleri;* this may be the result of hybridity. A few other specimens also show this, of which may be cited *Gillman* 1391 from Tanganyika, Lindi District, Njonga [Ngongo on EA sheet].

11. **S. dysophylla** *Benth.* in J.L.S. 1 : 103 (1857) ; F.T.A. 4 (1) : 533 (1903) ; Bruce & Lewis in K.B. 1956 : 273 (1956). Type : Portuguese East Africa, Delagoa Bay, *Forbes* 62 (K, holo. !)

Shrub or small much-branched tree, 2–12 m. high, with spreading branches and greyish to dark brown branchlets bearing round pale lenticels ; young branchlets glabrous to pubescent, often densely covered with conspicuous elongate lenticels. Leaves subsessile, ± membranous to coriaceous, obovate to elliptic, very rarely more than twice as long as broad, 1·5–7 cm. long, 1–5 cm. wide, rounded or subacute, cuneate or rounded at the base, pubescent to glabrous and shining. Cyme axillary, shortly pedunculate or subsessile, pubescent or glabrous. Flowers 4-merous. Calyx-lobes suborbicular, about 2 mm. long, ciliate. Corolla whitish or yellowish ; tube cylindric, about 3·5 mm. long, hirsute at the throat ; lobes very thick and cucullate, oblong, subacute. Anthers sessile in the corolla-throat. Ovary elongate-ovoid ; style hirsute at the base. Fruit 2·5–7 cm. diameter, globose ; rind hard, yellow, rarely black. Seeds about 7–12, tetrahedral, fairly thick, immersed in edible pulp.

subsp. **dysophylla** ; Bruce & Lewis in K.B. 1956 : 274 (1956)

Branchlets brown, thinly lenticellate ; lateral branchlets short, opposite and decussate, with congested leaf-scars. Leaves usually membranous, pubescent, at least on the nerves beneath ; upper surface often shining ; venation not very conspicuous. Fruit up to 7 cm. in diameter but usually smaller.

TANGANYIKA. Dodoma District : Meia Meia, 19 Aug. 1928 (sterile), *Greenway* 798 ! ; Mpwapwa, 27 Dec. 1931 (sterile), *H. E. Hornby* 428 ! ; Songea District : near Gumbiro, 26 Jan. 1956 (fl), *Milne-Redhead & Taylor* 8552 !
DISTR. **T4–6, 8** ; Portuguese East Africa southwards to Natal.
HAB. Deciduous bushland & woodland ; 900–1500 m.

SYN. *S. randiiformis* Baill. in Bull. Soc. Linn. Paris 1 : 246 (1880). Type: Portuguese East Africa, Delagoa Bay, *Forbes* 62 (?P, holo.)

subsp. **engleri** (*Gilg*) *Bruce & Lewis* in K.B. 1956 : 275 (1956). Type : Tanganyika, " Nyika steppe ", *Holst* 2420 (B, holo.†, HBG, K, iso. !)

Branchlets greyish, usually densely lenticellate, with opposite and decussate elongate lateral branchlets without congested leaf-scars. Leaves coriaceous, glabrous and shiny on both surfaces ; venation often conspicuous and reticulate. Fruits usually about 2·5–3 cm. in diameter, yellow or dull orange (rarely black).

KENYA. Mombasa District : Port Tudor, *MacNaughton* 90 *in F.D.* 2622 ! ; Kilifi District : near Roka, June 1937 (fl.), *Dale* 3803 ! ; Kwale District : between Samburu and Mackinnon Road, 31 Aug. 1953 (fr.), *Drummond & Hemsley* 4077 !

TANGANYIKA. Tanga District : 8 km. SE. of Ngomeni, 31 July 1953 (young fr.), *Drummond & Hemsley* 3570 ! ; Lushoto District : Makuyuni, June 1935 (young fl.), *Koritschoner* 1179 !

DISTR. **K**7 ; **T**3 ; Portuguese East Africa.

HAB. Bushland and open woodland ; 0–450 m.

SYN. *S. engleri* Gilg in E.J. 17 : 568 (1893) ; P.O.A. C : 310, t. 38 (1895) ; F.T.A. 4 (1) : 532 (1903) ; T.T.C.L. : 275 (1949)
S. wakefieldii Baker in K.B. 1895 : 98 (1895) ; F.T.A. 4 (1) : 532 (1903) ; T.S.K. : 126 (1936) ; T.T.C.L. : 276 (1949). Type : Kenya, Mombasa, *Wakefield* (K, holo. !)

12. **S. panganensis** *Gilg* in P.O.A. C : 311 (1895) ; F.T.A. 4 (1) : 526 (1903); E.J. 36 : 94 (1905) ; T.T.C.L. : 274 (1949) ; E. A. Bruce in K.B. 1956 : 154 (1956). Type : Tanganyika, Pangani, " Muhango," *Stuhlmann* 596 (B, holo.†, HBG, iso. ! K, photo.-iso. !)

Scandent or scrambling evergreen shrub ; young branchlets striate, usually crisped-pubescent and not conspicuously lenticellate ; older ones paler and lenticellate ; tendrils usually numerous, bifurcate. Leaves subsessile or shortly petiolate ; lamina usually coriaceous, variable in size and shape, ovate-lanceolate, ovate, elliptic, ovate-cordate or more rarely suborbicular, 1–6 cm. long, 1–3 cm. wide (the shorter and broader cordate leaves are at the base of lateral shoots and on the main stem), acute or shortly acuminate, often apiculate (rarely obtuse), cordate, cuneate, or rounded at the base, glabrous, except for a few hairs on the midrib beneath, strongly 3-nerved from the base, with a pair of fainter submarginal nerves ; venation prominent on both surfaces except for impressed midrib above. Cymes compound, divaricate, terminal and axillary, many-flowered, rather open ; peduncle and pedicels slender, rather long (6–14 mm. and 3–7 mm. respectively), pubescent with ascending brown hairs ; bracts and bracteoles narrowly lanceolate, acute. Flowers 5-merous, fragrant. Calyx-lobes ovate to oblong-ovate, obtuse or subacute, ciliate. Corolla cream-coloured or white ; tube campanulate, about 1·5 mm. long, pilose at the throat, scarcely half the length of the strap-shaped oblong glabrous subacute often reflexed corolla-lobes. Filaments 1·5–2 mm. long, inserted in the sinuses of the corolla-lobes ; anthers 1–1·5 mm. long, well-exserted. Style longer than the ovary, thinly pilose at the base. Fruit globose, 1·1–1·6 cm. in diameter, with a thin rind. Seeds (1–)3–5(–10), subtetrahedral, compressed, 5–8 mm. long.

KENYA. Kwale, Nov. 1928 (fl.), *R. M. Graham* 1581 ! ; Kilifi District : Kibarani, 12 Sept. 1936, *Swynnerton* 9 ! ; Kilifi, 5 Nov. 1937 (fl.), *Moggridge* 503 !

TANGANYIKA. Lushoto District : between Daluni & Mashewa, 26 Oct. 1935 (fl.), *Greenway* 4135 ! ; Tanga District : 8 km. SE. of Ngomeni, 31 July 1953 (fr.), *Drummond & Hemsley* 3572 ! ; Lindi District : Kwa Sikumbi, 16 June 1903, *Busse* 2909 !

DISTR. **K**7 ; **T**3, 6, 8 ; coastal region of Portuguese East Africa.

HAB. Lowland rain-forest and coastal evergreen bushland ; 50–500 m.

SYN. *S. guerkeana* Gilg in P.O.A. C : 311 (1895) ; Baker in F.T.A. 4 (1) : 521 (1903) ; E.J. 36 : 94 (1905) ; T.T.C.L. : 277 (1949) ; B.S.B.B. 85 : 27 (1952). Type : Tanganyika, Bagamoyo District, Rossako [Ruseko], *Stuhlmann* 8053 (B, holo.†, K, iso. !)
S. bicirrifera Dunkley in K.B. 1935 : 263 (1935) ; T.S.K. : 126 (1936) ; B.S.B.B. 85 : 27, fig. 8B (1952). Type : Kenya, Kilifi District, Arabuko, *R. M. Graham* 2290 (K, holo. !, EA, FHO, iso. !)

13. **S. angolensis** *Gilg* in E.J. 17 : 571 (1893) ; Hiern in Cat. Welw. Afr. Pl. 1 : 703 (1898) ; F.T.A. 4 (1) : 522 (1903) ; Lejeunia 11 : 67 (1947) ;

E. A. Bruce in K.B. 1956 : 157 (1956). Type : Angola, Cuanza Norte, Pungo Andongo, *Welwitsch* 6020 (B, holo.†, BM, K, iso. !)

A woody climber, scandent shrub or a tree with straggling branches, 5–12 m. high, with usually dense short tawny- or dark-pubescent and glabrescent branchlets often black or dark-coloured, and normally spreading ± at right angles and not obviously lenticellate ; tendrils simple, pubescent, coiled, often present on the young shoots. Leaves petiolate ; petiole short, transversely rugose, pubescent ; lamina coriaceous or subcoriaceous, very rarely submembranous, very variable in shape, oblong, oblanceolate-elliptic, ovate-lanceolate or broadly ovate, 2–9 cm. long, 1–5·5 cm. wide, usually rounded and apiculate, rarely emarginate or subacute at the apex, rounded, cuneate or more rarely subcordate at the base, 3–5-nerved from or just above the base, with an obvious marginal nerve and prominent midrib beneath, often thinly ciliate, pubescent on the midrib particularly towards the base, otherwise glabrous ; venation often ± inconspicuous above, more prominent beneath. Cymes axillary, paniculate, variable in size, 1–5 cm. long, 0·6–1 cm. broad ; lateral branches usually 3-flowered ; peduncles pubescent. Flowers 4–5-merous. Calyx-lobes ovate-deltoid, ciliate, otherwise glabrous. Corolla white, fading to yellow, or brown or terra-cotta, subrotate ; lobes oblong-ovate, about 1·5 mm. long, thinly pilose within, otherwise glabrous ; tube very short. Stamens inserted in the sinuses of the corolla-lobes, as long as the ovate glabrous anthers. Ovary ovoid, glabrous ; style short, glabrous. Fruit subglobose, 1–1·5 cm. in diameter, soft and fleshy, orange. Seed solitary, ellipsoid, 1 cm. long, 0·8 cm. in diameter, not compressed.

Tanganyika. Buha District : Mukugwe River, 48 km. S. of Kibondo, July 1951 (fl.), *Eggeling* 6212 ! ; Ulanga District : Mahenge, Mar. 1932 (fl.), *Schlieben* 1932 !
Zanzibar. Pemba, Chake-Chake, Mar. 1928, *Vaughan* 297 ! & 18 Aug. 1929 (fl.), *Vaughan* 521 ! ; Wesha, 17 Aug. 1929, *Vaughan* 529 !
Distr. **T**4, 6 ; **P** ; Nigeria eastwards through the Cameroons & northern Belgian Congo, also in Northern Rhodesia & Angola.
Hab. Riverine forest & coastal evergreen bushland ; 0–1500 m.

Syn. *S. angolensis* var. *tanganykae* Duvign. in Lejeunia 11 : 70 (1947). Type : Tanganyika, Mahenge, *Schlieben* 1932 (BR, holo. !, BM. iso. !)

Note. The above conception of *S. angolensis* includes *S. lacourtiana* De Wild., *S. bequaertii* De Wild., *S. likimiensis* De Wild., *S. mongonda* De Wild., *S. nauphylla* Duvign. and *S. tuvungasala* Duvign. from the Belgian Congo, and *S. cinnabarina* Hutch. & Dalz. from French Equatorial Africa.

14. **S. decussata** (*Pappe*) *Gilg* in E.J. 28 : 121 (1899) ; Verdoorn in Bothalia 3 : 587–8, fig. 2 (1939) ; E. A. Bruce in K.B. 1956 : 156 (1956). Type : South Africa, Cape Province, Bathurst, Kowie River, *Atherstone* (K, iso. !)

Erect bushy glabrous (apart from the flowers) shrub or small tree, 3–5 m. high, with smooth grey bark and neither spines nor tendrils. Branches opposite and decussate, ultimate twigs ± ochre-coloured, older ones dark-coloured and lenticellate, both with persistent swollen petiole-bases. Leaves shortly petiolate ; lamina subcoriaceous, obovate, oblong-obovate or elliptic, 1·5–4·5 cm. long, 0·8–3 cm. wide, usually rounded but sometimes emarginate or acute, cuneate at the base, usually dull, 3-nerved from or just above the base, with a fainter submarginal and marginal nerve along the somewhat inrolled margin ; venation not conspicuous above. Cymes axillary, rather numerous, laxly 3–9-flowered ; bracts and bracteoles conspicuous ; peduncles 4–10 mm. ; pedicels 3–4 mm. long. Flowers 4–5-merous, scented, about 4 mm. long. Calyx-lobes ovate, shortly ciliate, rounded or subacute. Corolla white or cream-coloured ; tube usually shortly campanulate, sometimes thinly pilose in the throat, otherwise glabrous ; lobes as long as or longer than the tube, oblong to oblong-deltoid, pilose inside. Stamens inserted in sinuses of the corolla-lobes, very shortly exserted. Style longer than the ovary, both glabrous. Fruit subglobose, 7–17 mm. in diameter, with a firm rind, yellow-orange. Seed solitary, ellipsoid, about 1 cm. long, slightly laterally compressed, with a depression (hilum) on one face.

KENYA. Kitui District : Nzui, 19 Jan. 1943 (fr.), *Bally* 1939 ! & Ikutha, 2 Nov. 1946 (fl.), *Edwards* 56 ! ; Teita District : Mt. Kasigao, Oct. 1938 (fl.), *Joana in C.M.* 8853 ! & Mbuyuni [between Ndi & Tsavo], 18 Nov. 1893, *Scott Elliot* 6235 !

DISTR. K4, 7 ; also in Portuguese East Africa and Southern Rhodesia and southwards to eastern Transvaal, Natal & eastern Cape Province in South Africa.

HAB. Deciduous bushland, often near rocks ; 500–1500 m.

SYN. *Atherstonea decussata* Pappe, Sylv. Cap., ed. 2 : 29 (1862)

NOTE. This has been called *S. atherstonei* Harv. in South Africa, but the name is illegitimate.

15. **S. boonei** *De Wild.* in B.J.B.B. 5 : 45 (1915) ; Duvign. in B.S.B.B. 85 : 32 (1952). Type : Belgian Congo, " Nala", *Boone* 83 (BR, syn. !)

Scandent shrub without spines. Ultimate branchlets brown-puberulous, often dark-coloured ; older branchlets pale brown, striate ; lenticels not or scarcely visible ; tendrils occasional, simple. Leaves petiolate ; petiole 4–7 mm. long ; blade membranous, elliptic, ovate or narrowly obovate, 2·5–10 cm. long, 0·9–4·5 cm. wide, abruptly longly acuminate, cuneate or more rarely rounded at the base, dull-surfaced, glabrous except for a few brown hairs at base of midrib beneath, strongly 3-nerved from the base and sometimes with a fainter submarginal pair ; tertiary venation close, spreading ± at right angles to the midrib ; nerves all prominent except for impressed midrib above. Cymes axillary and terminal, lax, spreading, pedunculate, 1·5–2·5 cm. long, bracteate, thinly pubescent with brown hairs. Flowers 5-merous, bracteolate. Calyx-lobes ovate to broadly ovate, 1·5 mm. long, 1 mm. broad, rounded to subacute, ciliate. Corolla white or greenish ; tube about 1–2 mm. long, glabrous ; lobes oblong, 2·5–3 mm. long, densely pilose within. Stamens inserted in the sinuses of the corolla-lobes, well exserted ; filaments about 1·2–1·7 mm. long ; anthers oblong, about 1 mm. long, glabrous. Ovary ovoid, about 1 mm. long, pilose ; style about 2·5 mm. long, pilose. Fruit not seen.

UGANDA. Mengo District : Bajo, June 1917 (fl.), *Dummer* 3218 ! & Mar. 1916 (fl.), *Dummer* 2771 !

DISTR. U4 ; Cameroons & northern Belgian Congo.

HAB. Lowland rain-forest ; 1300 m.

NOTE. *A. S. Thomas* 21 (from Uganda, Masaka District, Bugala, in rain-forest on clay soil) most probably belongs here. It is a vigorous sterile shoot of a scandent shrub and has long simple tendrils.

16. **S. kasengaënsis** *De Wild.* in B.J.B.B. 5 : 46 (1915) ; Duvign. in B.S.B.B. 85 : 27, fig. 8c (1952). Type : Belgian Congo, Katanga, Kasenga, *Bequaert* 237 (BR, holo. !, K, photo. !)

Woody climber up to 12 m. high. Stems striate, velutinous when young, with few and inconspicuous lenticels ; branchlets spreading at a wide angle ; tendrils bifurcate. Leaves shortly petiolate ; lamina coriaceous, narrowly ovate, rarely elliptic, up to 5 cm. long, 2·5 cm. wide (very much reduced on whippy climbing branches), acute or acuminate, rounded at the base, glabrous above, pubescent on the midrib beneath, ciliate when young, shiny, 3-nerved from the base, more strongly so beneath ; tertiary venation reticulate, more prominent beneath. Cymes compound, congested, axillary and terminal, about 1·5 cm. long, pubescent. Flowers 5-merous. Calyx-lobes ovate, ciliate. Corolla white; lobes about as long as the tube, oblong-lanceolate, scarcely acute, about 3 mm. long, glabrous or nearly so; tube subcylindrical, densely pilose within. Stamens inserted in the sinuses of the corolla-lobes, exserted; anthers about 1 mm. long, a little shorter than the filaments. Ovary pubescent; style about 2 mm. long, glabrous. Fruit globose, up to 4 cm. in diameter; rind hard and thin. Seeds up to 30, variously compressed, up to 13 mm. long, 5 mm. wide.

TANGANYIKA. Ufipa District: Kasanga, 16 June 1957 (fr.), *Richards* 10125! & 19 June 1957 (fr.), *Richards* 10167! & 30 Mar. 1959 (fl.), *Richards* 12299!
DISTR. **T4**; also in the Belgian Congo (Katanga) & Northern Rhodesia.
HAB. Tall deciduous thicket; 900 m.

17. **S. scheffleri** *Gilg & Busse* in E.J. 36: 95 (1905); T.T.C.L.: 276 (1949); E. A. Bruce in K.B. 1956: 153 (1956). Types: Tanganyika, E. Usambara Mts., Nderema, *Scheffler* 78 (B, syn.†) & Bomole Mt. near Amani, *Busse* 2200 (B, syn.†) & near Amani, *Warnecke* 389 (B, syn.†, EA, isosyn.!) & Kilwa District, Matumbi Mt., *Busse* 3124 (B, syn.†, EA, isosyn.!)

Woody climber up to 25 m. high. Stems striate, glabrous or sparsely pubescent with brown hairs, not or rarely lenticellate; branchlets spreading at right angles; tendrils bifurcate. Leaves shortly petiolate, rather variable in size; lamina coriaceous or subcoriaceous, ovate, obovate or elliptic, 4–10 cm. long, 2–5 cm. wide, acutely and sometimes shortly and abruptly acuminate at the apex, cuneate, rounded or more rarely subcordate at the base, glabrous or pubescent on the midrib beneath, not or scarcely shiny, strongly 3-nerved from the base and with a fainter submarginal pair; nerves prominent except for the impressed midrib above; tertiary venation reticulate. Cymes compound, many-flowered, congested or spreading, axillary, 2–6 cm. long, pubescent. Flowers scented, 5-merous. Calyx-lobes ovate to suborbicular, ciliate. Corolla white or cream-coloured; lobes as long as or slightly longer than the tube, oblong, subacute, 2·5–4 mm. long, glabrous, erect or reflexed; tube subcylindrical, densely pilose within. Stamens inserted in the sinuses of the corolla-lobes, exserted; anthers 1·7–2 mm. long, as long as or shorter than the filaments. Ovary pilose; style 4–6 mm. long, pilose at least at the base. Fruit globose, 2–3·5 cm. in diameter, with a woody rind. Seeds 8–?15, compressed, 1 cm. long, 1·5 cm. wide. Fig. 4/3, 4, p. 22; Fig. 5/4–6, p. 24.

var. **scheffleri**; E. A. Bruce in K.B. 1956: 153 (1956)

Inflorescence usually congested, 2–3 cm. long; corolla-lobes not reflexed; stamens not conspicuously exserted; style glabrous in the upper part.

KENYA. Kwale District: Shimba Hills, Makadara, Aug. 1929 (fl.), *R. M. Graham* 2045! & Cha Simba Forest, 2 Feb. 1953 (young fl.), *Drummond & Hemsley* 1094; without locality, *C. F. Elliott* 327!
TANGANYIKA. Lushoto District: E. Usambara Mts., Amani, 16 Nov. 1928 (fr.), *Greenway* 999! & 18 Sept. 1908, *Braun in A.H.* 7690!; Tanga District: E. Usambara Mts., Ngua Estate, 19 July 1953 (fl. & fr.), *Drummond & Hemsley* 3354!; Buha District: E. of R. Mlagarasi Ferry, Feb. 1955, *Procter* 385!
DISTR. **K7**; **T3**, 4, 8; possibly **Z** (see note); not known elsewhere.
HAB. Lowland rain-forest and its secondary growth; 300–1000 m.

NOTE. *Vaughan* 1332 from Pangajuu, Zanzibar Is., probably belongs here, mainly because of its bifurcate tendrils; being a sterile specimen, it cannot be more accurately placed.
The other variety of this species, var. *expansa* E. A. Bruce, which has a more diffuse inflorescence, 3–6 cm. long, occurs in the Cameroons, the Belgian Congo and Angola.

18. **S. xylophylla** *Gilg* in E.J. 28 : 122 (1899) ; F.T.A. 4 (1) : 526 (1903) ; T.T.C.L. : 277 (1949) ; E. A. Bruce in K.B. 1956 : 155 (1956). Type : Tanganyika, Uzaramo District, Kikulu, *Stuhlmann* (B, holo.†) and 16 km. S. of Dar es Salaam, *Vaughan* 2763 (BM, neo. !, EA, isoneo.)

Glabrous shrub with neither spines nor tendrils. Branchlets yellowish, smooth, lenticellate, subtetragonal, with four ± conspicuous raised ridges. Leaves subsessile or shortly petiolate ; lamina yellowish, coriaceous, obovate to broadly elliptic, 6–8 cm. long, 4–6 cm. wide, rounded and then shortly acuminate at the apex, cuneate or more rarely rounded at the base, ± shining above, dull below, 5–nerved from the base and with a marginal pair of nerves along the somewhat inrolled margins ; tertiary nerves prominent and conspicuously reticulate above, not so beneath. Cymes compound, axillary and terminal, pedunculate ; peduncle 2–22 mm. long. Flowers 4-merous, pedicellate, up to 6 per cyme, 2–5 mm. long and 4 mm. in diameter when open. Calyx-lobes broadly ovate-orbicular, ciliate. Corolla subrotate ; lobes ovate, about 2 mm. long, much longer than the tube, acute, pubescent within in lower part, glabrous outside. Stamens inserted in the sinuses of the corolla-lobes, exserted. Ovary globose ; style short. Fruit subglobose, about 1·5 cm. in diameter with a firm rind, orange. Seeds 3, subtetragonal, slightly compressed, about 10 mm. long, 8 mm. wide.

TANGANYIKA. Uzaramo District : 16 km. from Dar es Salaam on Utete road, 24 Feb. 1939 (fr.), *Vaughan* 2763 !
DISTR. **T**?3, 6 ; not known elsewhere.
HAB. Riverine forest ; ± 100 m.

NOTE. As the only specimen seen of this species lacks mature flowers its relationship to the other species is uncertain. The fruits resemble those of *S. panganensis*; the branchlets and leaves are more reminiscent of *S. henningsii*. A specimen from Tanganyika, Handeni District, Sindeni Hill (*B. D. Burtt* 5419), probably belongs here, but the immaturity of its inflorescences denies certainty.

Gatherings of this species in flower would be very valuable. The yellow-green leaves and the small few-seeded orange fruits should make it easy to recognize.

19. **S. henningsii** *Gilg* in E.J. 17 : 569 (1893) ; Fl. Cap. 4 (1) : 1052 (1909) ; Bothalia 3 : 583–588, fig. 1 (1939) ; E. A. Bruce in K.B. 1955 : 127 (1955). Types : South Africa, Pondoland, *Bachmann* 1745 (B, syn.†) & *Beyrich* 1 (B, syn.†) and Cape Province, Peri Forest, near King Williamstown, *Scott Elliot* 979 (K, neo. !)

Small, erect, much-branched tree or shrub, 2–10 m. high, with a spreading rounded crown and neither spines nor tendrils. Branchlets pale, ashy or straw-coloured, ± swollen at the nodes and with persistent petiole-bases ; young branchlets subquadrangular ; lenticels few and inconspicuous. Leaves subsessile, usually rather congested and very variable in size and shape even on the same branchlet, coriaceous or subcoriaceous, broadly ovate, ovate-lanceolate, elliptic or oblong-elliptic, 2–9 cm. long, 1–6 cm. wide, rounded, acute or acuminate at the apex, rounded, cuneate or rarely subcordate at the base, usually shiny above, 3-nerved from the base and often with 1, rarely 2, submarginal nerves on each side ; tertiary nerves usually very prominent on both surfaces. Cymes usually compound, axillary and terminal, subsessile or pedunculate ; peduncle 0·2–2 cm. long. Flowers fragrant, sometimes cleistogamous, 4–6-merous, sessile or subsessile, often densely clustered, 2–2·5 mm. long and 4 mm. in diameter when open. Calyx-lobes broadly ovate-orbicular, ciliate. Corolla yellow or cream-coloured, subrotate ; lobes ovate-deltoid, about 1·5 mm. long, subequal to or longer than the tube, acute or subacute, pubescent within (particularly at the base), more rarely glabrous, usually glabrous outside. Stamens inserted in the sinuses of the corolla-lobes, just exserted. Ovary globose ; style short.

Fruit globose or ovoid, 8–14 mm. long, 6–11 mm. broad, with a thin rind, red or brown. Seeds usually like coffee-beans with an elongated ridged groove down one side, ellipsoid or elongate-ellipsoid, 9–10 mm. long, 6 mm. broad.

UGANDA. Bunyoro District : Bukumi, May 1933 (fl.), *Eggeling* 1228 *in F.D.* 1331 ! ; Busoga District : Bukoli, Siavona Hill, 26 Mar. 1953 (fl.), *Wood* 667 ! ; Mengo District : Buruli, near Nakasongola, July 1940 (fr.), *Eggeling* 3976 !

KENYA. Northern Frontier Province : Moyale, 23 Apr. 1952, *Gillett* 12914 ! ; Nairobi District : Mbagathi road, May 1934 (fl.), *van Someren* 3249 *in C.M.* 6260 ! ; Kwale District : 8 km. E. of Mackinnon Road, 9 Sept. 1953 (fl.), *Drummond & Hemsley* 4226 !

TANGANYIKA. Lushoto District : W. Usambara Mts., Kwebao Forest, 18 Aug. 1953 (fl.), *G. R. Williams* 498 ! ; Shinyanga District : Mantini Hills, 26 Mar. 1932 (fr.), *B. D. Burtt* 3736 ! ; Lindi District : Sudi, 12 Dec. 1942 (fl.), *Gillman* 1137 !

DISTR. **U**1–4 ; **K**1–2, 4–7 ; **T**1, 3, 6, 8 ; Belgian Congo, Angola, the Sudan, Somaliland and southwards to eastern Cape Province, South Africa.

HAB. Upland and lowland rain-forest, semi-evergreen bushland, lowland dry evergreen forest & riverine forest ; 340–2000 m.

SYN. *S. holstii* Gilg in P.O.A. C : 310 (1895) ; F.T.A. 4 (1) : 529 (1903) ; Fl. Pl. Sudan 2 : 383 (1952). Types : Tanganyika, E. Usambara Mts., Mashewa, *Holst* 8833a (B, holo.†) & Pare District, S. Pare Mts., between Chome and Vudea, *Greenway* 6562 (K, neo. !, EA, isoneo. !)

S. procera Gilg & Busse in E.J. 36 : 97 (1905) ; T.T.C.L. : 276 (1949). Types : Tanganyika, Lindi District, island in Lake Lutamba, *Busse* 2506, 2511, 2511a & 2516 (B, syn.†, EA, HBG, isosyn. !) & between Muera [Rondo] & Noto Plateaux, Kwa-Sikumbi, *Busse* 2903 (B, syn.†)

S. albersii Gilg in E.J. 36 : 99 (1905) ; T.T.C.L. : 277 (1949). Type : Tanganyika, W. Usambara Mts., Kwai, *Albers* 380 (B, holo.†, EA, iso. !)

S. elliottii Gilg in E.J. 36 : 99 (1905). Type : Kenya, Nairobi, *C. F. Elliott* 176 (B, holo.†, EA, K, iso. !)

S. myrcioïdes S. Moore in J.B. 45 : 52 (1907) ; I.T.U., ed. 2 : 169 (1952). Type : Uganda, Bunyoro District, Butiaba Plain, *Bagshawe* 841 (BM, holo. !)

S. reticulata Burtt-Davy & Honoré in K.B. 1932 : 270 (1932) ; T.S.K. : 125 (1936) ; T.T.C.L. : 276 (1949). Type : Kenya, unlocalized, *Conservator of Forests* 40 (K, holo. !)

S. holstii Gilg var. *procera* (Gilg & Busse) Duvign. in Bull. Inst. Roy. Col. Belge 20 : 587 (1949)

S. holstii Gilg var. *reticulata* (Burtt-Davy & Honoré) Duvign. in Bull. Inst. Roy. Col. Belge 20 : 587 (1949)

NOTE. The above conception of *S. henningsii* includes also *S. pauciflora* Gilg and *S. sennensis* Baker from Portuguese East Africa and *S. ligustroïdes* Gossw. & Mend. from Angola.

At least in some localities *S. henningsii* reproduces cleistogamously.

20. **S. stuhlmannii** *Gilg* in E.J. 17 : 570 (1893) ; P.O.A. C : 310 (1895) ; F.T.A. 4 (1) : 529 (1903) ; T.T.C.L. : 276 (1949) ; B.S.B.B. 86 : 108 (1953). Types : Tanganyika, Mwanza, *Stuhlmann* 4178 (B, syn.†) & Portuguese East Africa, Tete Province, R. Zambezi opposite Chiramba [cited as Shinamba], *Kirk* (K, syn. !)

Tree 6–18 m. high, with smooth grey bark, glabrous except for the flowers. Branchlets unarmed and without tendrils, pale brown, usually conspicuously lenticellate and twiggy, with protruding persistent cup-like petiole-bases. Leaves shortly petiolate ; lamina membranous to subcoriaceous, oblong-ovate to broadly ovate, 6–16 cm. long, 2–10 cm. wide, acute or acuminate, rounded or cuneate at the base, dull or more rarely shining above, dull beneath ; main pair of lateral nerves arising 5–7 mm. above the lower pair which arise near the base ; both pairs running ± parallel to the margin ; secondary nerves arising at a wide angle with the midrib ; tertiary venation reticulate, prominent beneath, less so above. Cymes axillary from the basal pair of leaves of the young shoot, sometimes appearing falsely terminal on the old. Flowers violet-scented, appearing before or with the young leaves ; pedicels and peduncles slender. Calyx-lobes oblong to ovate, glabrous.

Corolla white ; lobes oblong, thin, cucullate, usually longer than the tube, pilose within towards the base. Anthers glabrous, subsessile or shortly filamented, inserted at the mouth of the corolla-tube. Style long, ± thick, exserted, gradually emerging from the glabrous ovary. Fruit purple-black, cherry-like, 1·3–1·8 cm. in diameter with a thin rind, eaten by birds but poisonous to man. Seed solitary, large, compressed-globose, about 1·6 cm. diameter.

TANGANYIKA. Mwanza District : Kissesa, 5 Aug. 1951 (fr.), *Tanner* 388 ! ; Shinyanga, *Koritschoner* 1809 ! ; Mpwapwa, 20 Dec. 1933 (fl.), *Hornby* 573 !

DISTR. **T**1, 2, 4–8 ; southwards to Bechuanaland and the northern Transvaal.

HAB. *Combretum* woodland, *Brachystegia* woodland, semi-evergreen bushland, often on termite-hills ; 350–1600 m.

SYN. *S. heterodoxa* Gilg in E.J. 28 : 118 (1899) ; F.T.A. 4 (1) : 530 (1903) ; T.T.C.L. : 275 (1949). Type : Tanganyika, " Uhehe, Muhinde Steppe," *Goetze* 519 (B, holo.†, K, iso. !)

21. **S. usambarensis** *Gilg* in P.O.A. C : 311 (1895) ; F.T.A. 4 (1) : 526 (1903) ; T.S.K. : 125 (1936) ; T.T.C.L. : 276 (1949) ; E. A. Bruce in K.B. 1955 : 627 (1956). Type : Tanganyika, E. Usambara Mts., Mashewa, *Holst* 3582 (B, holo.†, HBG, K, iso. !)

Small unarmed tree or shrub, 3–12 m. high, sometimes sarmentose with long thin branches. Branchlets dark, spreading, glabrous, usually conspicuously lenticellate, ultimate ones often covered with a pale skin, which later splits and peels off ; tendrils occasional, simple. Leaves petiolate ; petiole 3–8 mm. long, often curved ; lamina thinly coriaceous, sometimes folded longitudinally, ovate, ovate-lanceolate or lanceolate, 4–7 cm. long, 1·3–3·8 cm. wide, fairly longly acuminate, cuneate or rounded at the base, glabrous ; lateral nerves in two pairs, arising at and up to 5 mm. above the base ; tertiary venation usually reticulate beneath. Cymes axillary, glabrous, about 1 cm. long, with deltoid bracts and bracteoles. Flowers 4–5-merous. Calyx-lobes ovate, acute or subacute, glabrous or minutely pubescent. Corolla white or yellow ; lobes ± oblong, about 2–3 mm. long, glabrous, subequal to or slightly longer than the campanulate tube, which is thinly pilose in the throat. Stamens inserted in the sinuses of the corolla-lobes ; anthers exserted. Ovary ovoid, glabrous; style rather long. Fruit globose, 1·0–1·8 cm. in diameter, sometimes shortly stipitate within the calyx, which, together with the bracts is persistent, ± fleshy, orange. Seed solitary, subglobose, slightly depressed with a central hilum, about 1 cm. long.

UGANDA. Karamoja District : Kaabong, 7 June 1942 (fl.), *Dale* 280 ! ; Toro District : Kibale Forest, 4 Oct. 1905 (fr.), *Dawe* 531 ! ; Busoga District : Bwenda, Aug. 1940, *Dale* 151 !

KENYA. Nairobi, 13 Mar. 1911 (fl.), *Battiscombe* 464 ! & 20 Mar. 1932, *van Someren* 1810 *in C.M.* 4781 ; Machakos District: about 11 km. SE. of Machakos, 3 Nov. 1953, *Hemming* 223!

TANGANYIKA. Lushoto District : E. Usambara Mts., Mangubu to Misoswe, 14 Feb. 1931, *Greenway* 2897 ! & about 1·5 km. E. of Mashewa, 29 June 1953, *Drummond & Hemsley* 3099 ! ; Tanga District : 8 km. SE. of Ngomeni, 2 Aug. 1953, *Drummond & Hemsley* 3607 !

DISTR. **U**1–3 ; **K**4, 6 ; **T**3, 5 ; Portuguese East Africa, Southern Rhodesia, the Transvaal & Natal.

HAB. Upland and lowland rain-forests, semi-evergreen and coastal evergreen bushlands ; 75–2000 m.

SYN. *S. micans* S. Moore in J.L.S. 40 : 146 (1911) ; Verdoorn in Bothalia 3 : 587 (1939). Type : Southern Rhodesia, Chirinda Forest, *Swynnerton* 125 (BM, holo., K, iso. !)

Imperfectly known species

Some species, which cannot be placed with certainty, though their affinity is fairly sure, have been either entered in synonymy with an interrogation mark or mentioned in a note under the species to which they are believed to belong. The following East African species cannot be placed even in this way :—

18. **S. cerasifera** *Gilg* in P.O.A. C : 311 (1895) ; F.T.A. 4 (1) : 531 (1903) ; T.T.C.L. : 277 (1949). Type : Tanganyika, coastal region, unlocalized, *Stuhlmann* 6089 (B, holo.†)

An erect shrub, without tendrils. Leaves ovate-oblong, 3–4·5 cm. long, 1·3–1·6 cm. broad, equally attenuate to base and apex, glabrous, extreme apex rounded ; venation inconspicuous. Inflorescences terminal. Fruit globose, 1·3–1·5 cm. in diameter, edible, 1-seeded.

19. **S. distichophylla** *Gilg* in P.O.A. C : 310 (1895) ; F.T.A. 4 (1) : 525 (1903) ; T.T.C.L. : 277 (1949). Type : Tanganyika, Biharamulo District, Kimoani [Kimwani] Plateau, *Stuhlmann* 3397 (B, holo.†)

Small, densely branched tree. Leaves ovate, 2–4 cm. long, 1·2–2 cm. broad, glabrous ; apex acute, apiculate. Inflorescences terminal. Fruit ovoid, 1·2–1·3 cm. long, 9–10 mm. in diameter, 1-seeded.

20. **S. pachyphylla** *Gilg & Busse* in E.J. 36 : 96 (1905) ; T.T.C.L. : 275 (1949). Type : Tanganyika, W. Usambara Mts., Kwai, *Eick* 332 (B, holo.†)

Erect shrub, without tendrils. Leaves coriaceous, broadly ovate, glabrous. Inflorescences axillary, pedunculate, many-flowered. Calyx-lobes rounded above, glabrous. (This may be a synonym of *S. innocua* Del.)

4. BUDDLEIA *

L., Sp. Pl. : 112 (1753) & Gen. Pl., ed. 5 : 51 (1754) ; Marquand in K.B. 1930 : 177 (1930)

Chilianthus Burch., Trav. 1 : 94 (1833)

Shrubs or small trees, hairy and glandular. Leaves rather thick, often wrinkled, frequently dentate or crenate ; stipules interpetiolar or reduced to a line. Cymes many-flowered, variously grouped into terminal and lateral panicles. Calyx 4-lobed or 4-fid. Corolla-tube ± cylindrical ; lobes spreading, imbricate in bud. Stamens 4, inserted in the mouth of the corolla, not exserted (except in *B. dysophylla*). Ovary 2-locular ; ovules numerous in rows on axile placentas. Capsule ± oblong, septicidally dehiscent and the valves also at least partially halving from above. Seeds small and numerous, sometimes winged or tailed ; endosperm fleshy ; embryo straight.

A genus occurring in tropical and South Africa, Madagascar and even more numerously in tropical Asia and America.

The three genera *Chilianthus* Burch., *Adenoplusia* Radlk. and *Nicodemia* Tenore are very closely related to *Buddleia* and opinions differ as to their correct status. We have not been able to examine the fruits of these, wherein lie their diagnostic characters. We have accepted both Radlkofer's ** and Phillips' ‡ judgment that *Chilianthus* should be included within *Buddleia*, and prevailing practice is also followed in treating *Adenoplusia* and *Nicodemia* (Madagascar and the Mascarene Is.) as distinct genera, even though they differ so little from *Buddleia* when in flower. Because of the possibility of confusion, two species of these genera have been included in the following key although *Nicodemia madagascariensis* (Lam.) Parker occurs only as a cultivated plant in East Africa.

* This spelling follows Rickett, see Taxon 4 : 187 (1955) ; Linnaeus's original spelling is *Buddleja*.

** See Abh. Nat. Ver. Bremen 8 : 410 (1883).

‡ See Journ. S. Afr. Bot. 12 : 114 (1946).

Also included in this key to *Buddleia* are two Asiatic species which are commonly cultivated, namely *B. davidii* Franch. (syn. *B. variabilis* Hemsl.) and *B. asiatica* Lam.

Corolla violet, lilac, mauve or partly so on a white ground, more than 4 mm. long ; leaves lanceolate or narrowly lanceolate, crenulate or finely serrate :
Leaves rounded or cordate below, crenate ; flowers in dense glomerules in a compound panicle . . . . . . 2. *B. salviifolia*
Leaves cuneate to attenuate below, margin serrate ; flowers in ± lax clusters in a simple or compound spike or raceme . *B. davidii*
Corolla orange, yellow or wholly white ; leaves various (never crenulate) :
Leaves lanceolate or narrowly elliptic, sessile or almost so, more than three times as long as broad ; flowers in spikes or racemes :
Corolla 4 mm. long or more, yellow or orange:
Leaf-margin partly serrate ; stem round 1. *B. polystachya*
Leaf-margin entire ; stem square . *Adenoplusia uluguruensis* (p. 41)
Corolla less than 4 mm. long, white . *B. asiatica*
Leaves elliptic or ovate, petiolate, less than three times as long as broad ; flowers either in racemes or panicles of cymules :
Corolla more than 4 mm. long, cylindric, usually yellow or orange ; stamens included ; leaf-margins entire or undulate:
Flowers in lax compound regularly tapering racemes of cymules (Fig. 7/3, p. 39) . . . . *Nicodemia madagascariensis* (pp. 35, 41)
Flowers in dense irregularly bunched panicles of cymose clusters . . 3. *B. pulchella*
Corolla less than 4 mm. long, campanulate, white ; stamens exserted ; leaf-margins dentate . . . . . . 4. *B. dysophylla*

1. **B. polystachya** *Fresen.* in Flora 21 : 605 (1838) ; F.T.A. 4 (1) : 515 (1903) ; Marquand in K.B. 1930 : 193 (1930) ; T.T.C.L. : 269 (1949) ; I.T.U., ed. 2 : 165, t. 8 (1952). Type : Ethiopia, unlocalized, *Rüppell* (FR, holo.)

Robust shrub, up to 5 m. tall, with pale brown bark ; young parts densely flocculent, glabrescent. Leaves shortly petiolate ; petiole up to 5 mm. long ; lamina up to 16 cm. long, 4 cm. wide, lanceolate, gradually narrowing above to an acute apex, convergent below, serrate at least in part. Inflorescence a compound raceme of short subglomerulate clusters ; lateral racemes up to 15 cm. long and less than 2 cm. in diameter ; pedicels obscure. Corolla reddish-orange ; corolla-lobes almost square, with rounded apex. Anthers subsessile, inserted just within the throat. Ovary narrowly ovoid, shortly hirsute. Fig. 6/1–5.

UGANDA. Karamoja District : Mt. Debasien, Jan. 1936, *Eggeling* 2748 ! ; Mbale District : Buginyanya, Sept. 1932, *A. S. Thomas* 462 !
KENYA. North Kavirondo District : Kakamega Forest, Nov. 1934, *Dale* 3388 ! ; Trans-Nzoia District : Hoey's Bridge, *Mainwaring* 39 ! ; Mt. Kenya, *Holyoak* 715 !

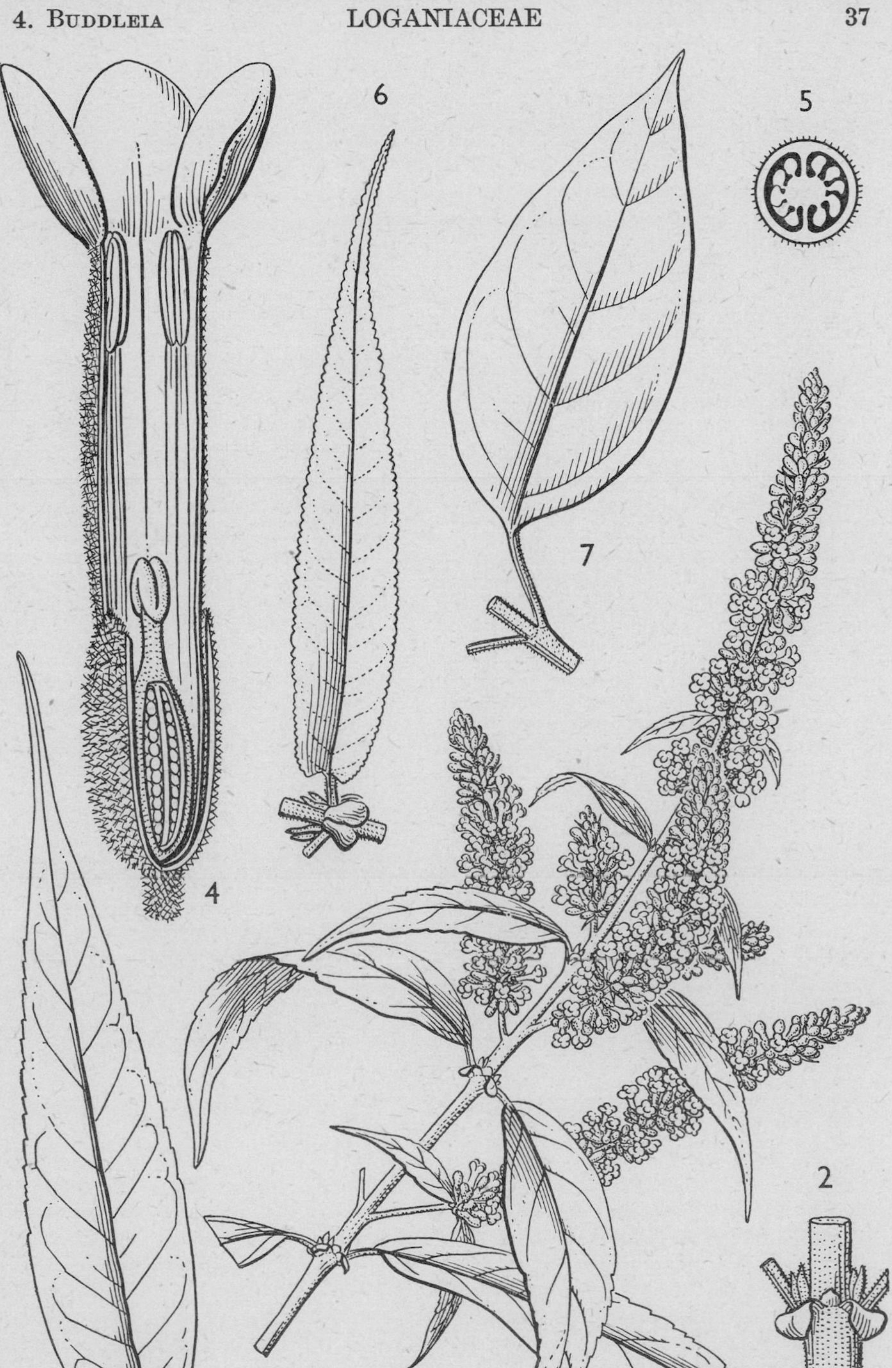

FIG. 6. *BUDDLEIA POLYSTACHYA*—**1**, flowering branch, × 1 ; **2**, node of stem showing stipules, × 3 ; **3**, leaf, × $\frac{2}{3}$; **4**, flower cut to show stamens and pistil, × 12 ; **5**, T.S. of ovary (diagrammatic) ; *B. SALVIIFOLIA*—**6**, leaf, × $\frac{2}{3}$; *B. PULCHELLA*—**7**, leaf, × $\frac{2}{3}$. 1, 2, 4, from *Linton* 174 ; 3, from *Albrechtsen* 1 ; 6, from *Carmichael* 251 ; 7, from *G. R. Williams* 62.

TANGANYIKA. Arusha District : W. slope of Mt. Meru, Sept. 1932, *B. D. Burtt* 4089 ! ; Mbulu District : S. slope of Mt. Hanang, Sept. 1930, *B. D. Burtt* 2370 !
DISTR. **U**1, 3 ; **K**2–6 ; **T**2 ; Somaliland, Ethiopia and Eritrea.
HAB. Upland grassland, margins and clearings of upland rain-forest ; 1000 to 2700 m.

SYN. *B. powellii* Kraenzl. in E.J. 50, Beibl. 111 : 34 (1913). Type : Kenya, unlocalized, *Powell* 73 (K, iso. !)

VARIATION. A small-leaved form from Eritrea has been described as var. *parvifolia* Marquand.

2. **B. salviifolia** (*L.*) *Lam.*, Encycl. 1 : 513 (1785) ; F.T.A. 4 (1) : 516 (1903) ; Marquand in K.B. 1930 : 198 (1930) ; T.S.K. : 125 (1936) ; T.T.C.L. : 270 (1949). Type : unlocalized, Herb Linnaeus (LINN, lecto. !)

Shrub or tree, up to 5 m. tall, with reddish-brown bark ; young parts tomentellous, glabrescent. Leaves shortly petiolate ; petiole about 3 mm. long ; lamina lanceolate or narrowly lanceolate, up to 10 cm. long, 2·5 cm. wide, shortly attenuate to an acute apex, rounded or cordate at the base, crenulate. Inflorescence a compound raceme of dense clusters of cymules ; lateral racemes up to 6 cm. long and 2·5 cm. in diameter ; pedicels obscure. Corolla lilac to mauve or violet, with darker or orange centres ; corolla-lobes very broadly oblong, with rounded apex. Anthers subsessile, inserted below the throat. Ovary almost spherical, densely hirsute above. Fig. 6/6, p. 37.

TANGANYIKA. Iringa District : W. Mufindi, Aug. 1933, *Greenway* 3482 ! ; Rungwe District : N. side Rungwe Mt., valley of Kiwira R., *B. D. Burtt* 6232 ! ; Njombe District : Elton Plateau, Nov. 1931, *R. M. Davies* 34 !
DISTR. **T**7 ; Nyasaland, Portuguese East Africa, Southern Rhodesia, South Africa.
HAB. Upland rain-forest margins, riverine forest in areas of upland grassland, persisting in secondary bushland ; 1450–2400 (2700) m.

SYN. *Lantana salviifolia* L., Syst. Nat., ed. 10 : 1116 (1759)
B. aurantiaco-maculata Gilg in E.J. 30 : 377 (1901) ; F.T.A. 4 (1) : 517 (1903). Type : Tanganyika, Mbeya District, Poroto Mts., Ngozi, *Goetze* 1129 (B, holo. †)

3. **B. pulchella** *N.E. Br.* in K.B. 1894 : 389 (1894) ; Marquand in K.B. 1930 : 190 (1930) ; T.T.C.L. : 270 (1949). Type : South Africa, Natal, *J. M. Wood*, cult. at Kew (K, holo. !)

Scandent shrub, up to 10 m. tall ; bark brown, stripping ; young parts densely floccose, glabrescent. Leaves petiolate ; petiole 0·5–2·0 cm. long ; lamina narrowly ovate to almost lanceolate or more rarely (and not in our area) hastate, 8–11(–15) cm. long, 3–5(–7) cm. wide, narrowing above to an acute or acuminate and minutely apiculate apex, rounded below to a cuneate-acuminate base ; margin entire, very shallowly undulate or very widely and bluntly dentate. Inflorescence a pyramidal compound raceme of clusters of cymules ; lateral racemes up to 7 cm. long and 3·5 cm. in diameter ; pedicels 1 mm. long. Corolla cream-coloured or yellow, orange in the throat ; tube brownish-cream ; corolla-lobes square, with truncate apex. Anthers subsessile, inserted just within the throat. Ovary ovoid, hirsute. Fig. 6/7, p. 37.

KENYA. Teita Hills, Ngangao, Sept. 1932, *Gardner* 2940 ! & 15 Sept. 1953, *Drummond & Hemsley* 4327 !
TANGANYIKA. Masai District : Loliondo, 16 km. W. of Klein's camp, Nov. 1953, *Tanner* 1815 ! ; Lushoto District : Shagai Forest, nr. Sunga, May 1953, *Drummond & Hemsley* 2568 !
DISTR. **K**7 ; **T**2, 3, ?6* ; Southern Rhodesia & South Africa.

* A specimen from the Uluguru Mts. (Morningside, 1200 m., 22 July 1958, *Carmichael* 667) has a denser inflorescence and smaller flowers, but may belong here. Further collections from the Uluguru Mts. are required.—Ed.

Fig. 7. Flowering branches, × ⅔, of 1, *BUDDLEIA DYSOPHYLLA*; 2, *ADENOPLUSIA ULUGURUENSIS*; 3, *NICODEMIA MADAGASCARIENSIS*. 1, from *Geilinger* 2, from *Schlieben* 2756; 3, from plant cult. at Kew.

HAB. Upland rain-forest and upland evergreen bushland, often persisting in secondary growth ; 1650–2000 m.

SYN. *B. usambarensis* Gilg in P.O.A. C : 313 (1895) ; F.T.A. 4 (1) : 516 (1903). Types : Tanganyika, Usambara Mts., Mshusa, *Holst* 8967 (B, syn.†, K, isosyn. !) & " Nyika ", *Holst* 3721 (B, syn.†)
?*B. oreophila* Gilg in E.J. 23 : 202 (1896) ; F.T.A. 4 (1) : 516 (1903) ; T.T.C.L. : 269 (1949). Type : Tanganyika, Uluguru Mts., Lukwangule, *Stuhlmann* 9101 (B, holo.†)
B. sp. in T.S.K. : 125 (1936)

NOTE. No material has been seen of *B. oreophila* Gilg but judging from the description it is a variant of this species. In South Africa this species has been known as *B. woodii* Gilg.

VARIATION. Specimens with their leaf-margins more irregularly wavy than can properly be called " undulate " are known from the more southern African territories ; they are mostly from cultivated or young plants.

4. **B. dysophylla** (*Benth.*) *Radlk.* in Abh. Nat. Ver. Bremen 8 : 410 (1883) ; Phillips in Journ. S. Afr. Bot. 12 : 114 (1946). Type : South Africa, Cape Province, Uitenhage, *Ecklon* (K, holo. !)

Climbing shrub, up to 3 m. tall, with rather shining brown bark ; young parts densely yellow-brown floccose. Leaves petiolate ; petiole up to 1 cm. long, sometimes appearing longer when including wings from the lamina ; lamina ovate or narrowly ovate, up to 11 cm. long, 5 cm. wide, narrowing above to an acute apex, truncate or rounded below to a cuneate base, especially in older leaves, dentate. Inflorescence a lateral foliaceous doubly compound raceme of ± distant cymose clusters, up to 15 cm. long, and about 7 cm. in diameter ; pedicels obscure. Corolla white; lobes oblong, rounded above. Stamens inserted at the mouth of the corolla-tube and ± longly exserted. Ovary subspherical, densely floccose. Capsule 2 mm. long, pubescent. Fig. 7/1, p. 39.

TANGANYIKA. Mbeya District : Poroto Mts., Sept. 1954, *F. G. Smith* 1279 ! ; Iringa District : nr. Dabaga, Sept. 1932, *Geilinger* ! ; Njombe District : Lupembe, Aug. 1931, *Schlieben* 1112A !
DISTR. T7 ; South Africa.
HAB. Upland rain-forest ; 1600–1900 m.

SYN. *Nuxia dysophylla* Benth. in Comp. Bot. Mag. 2 : 60 (1836)
Chilianthus dysophyllus (Benth.) DC., Prodr. 10 : 436 (1846) ; Fl. Cap. 4 (1) : 1045 (1909) ; T.T.C.L. : 270 (1949)

NOTE. It is not yet possible to be sure that this species is correctly placed here. The stamens are similar to those of *Nuxia* and quite unlike those characteristic of *Buddleia*. It may, therefore, have to be treated as a separate genus.

5. ADENOPLUSIA

Radlk. in Abh. Nat. Ver. Bremen 8 : 461 (1883)

Shrubs with angled branches and usually large interpetiolar stipules. Leaves elliptic or narrowly elliptic, papery, often glandular. Inflorescences of few-flowered whorls or clusters of usually simple single axillary racemes or spikes. Flowers similar to those of *Buddleia*. Fruit a " drupe "*.

A genus of several species in Madagascar but with only a single species on the mainland of Africa. The Madagascar species are very closely interrelated but ours, as well as being

* To describe the fruits of this genus as drupes is probably not correct. In one species, for example (*A. axillaris* Radlk., as collected by Baron in Madagascar), the dried fruits have only thin walls, are longitudinally sutured like the capsules of *Buddleia* and the numerous seeds form a mass quite free from the wall. If they are not fleshy capsules (not being ultimately dehiscent) they should be called berries.

far removed geographically, is distinguishable from all of them by its leaves, flowers and fruits.

When the subfamily *Buddleioïdeae* is analysed in full it is probable that *Adenoplusia* and *Nicodemia* Tenore will be reduced to subgenera of *Buddleia*.

A. uluguruensis *Melch.* in N.B.G.B. 12 : 203 (1934) ; T.T.C.L. : 268 (1949). Type : Tanganyika, NW. Uluguru Mts., *Schlieben* 2756. (B, holo.†, BM, iso. !, K, drawing of iso. !)

Shrub with pale brown 4-angled slightly winged branches, glabrescent ; stipules large, foliaceous, about 0·6 cm. long, 1·0 cm. wide. Leaves sessile, narrowly elliptic, up to 16 cm. long, 4 cm. wide, acuminate-caudate, cuneate-attenuate below to a 0·5 mm. broad base, entire, glabrous above, glabrous but glandular beneath. Racemes about 5 cm. long ; flowers sessile in spaced whorls. Corolla yellow, the tube more than twice the length of the calyx. Ovary ovoid, bilocular, glandular. Fruit 4 mm. long, 3 mm. broad, partly enclosed by the persistent calyx. Fig. 7/2, p. 39.

TANGANYIKA. Morogoro District : NW. Uluguru Mts., Sept. 1932, *Schlieben* 2756 !
DISTR. T6 ; not known elsewhere.
HAB. Upland rain-forest, in open spaces ; 1380 m.

6. NUXIA

Lam., Illustr. 1 : 295, t. 71 (1791)

Shrubs or small trees, glandular and ± hairy. Leaves petiolate ; stipules reduced to a mere line. Flowers 4-merous, small and numerous, in cymose panicles. Calyx glandular, often externally ; lobes valvate, persistent. Corolla-tube ± cylindrical, circumscissile just above the persistent base ; lobes imbricate in aestivation. Stamens inserted just below mouth of corolla-tube, longly exserted ; filaments filiform. Ovary 2-locular ; ovules numerous in rows on axil placentas. Capsule ovoid or obloid, septicidally dehiscent and the valves halving above. Seeds small and numerous ; endosperm fleshy ; embryo straight.

A genus of tropical and South Africa, and Madagascar.

This treatment returns to the generic concept of Bentham * and of J. G. Baker ** and agrees with that of P. Jovet‡, the mainland species (*Lachnopylis* sensu C. A. Smith §) being considered congeneric with the Mascarene type-species, *Nuxia verticillata* Lam. The difference in the inflorescence of this latter is insufficient to distinguish it generically, especially in view of the inflorescence-variation which is accepted within the allied genus *Buddleia*.

Leaves usually distantly dentate, bluntly serrate or, at least, sinuate, rarely entire ; calyx 4–5 mm. long, distinctly 5-lobed ; corolla-lobes rounded above :
- Leaves acute or acuminate ; ovary and fruit glabrous ; fruit almost twice the length of the calyx . . . . 1. *N. floribunda*
- Leaves rounded above (and often mucronulate) ; ovary and fruit hirsute ; fruit scarcely exceeding the calyx in length . . . . . 2. *N. oppositifolia*

Leaves entire ; calyx about 5 mm. long ; lobes often failing to separate regularly into 5 lobes ; corolla-lobes narrowing acutely above . . . 3. *N. congesta*

* See DC., Prodr. 10 : 434 (1846).
** See F.T.A. 4 (1) : 511 (1903).
‡ See Bull. Hist. Nat. Toulouse 82 : 4 (1948).
§ See K.B. 1930 : 10 (1930).

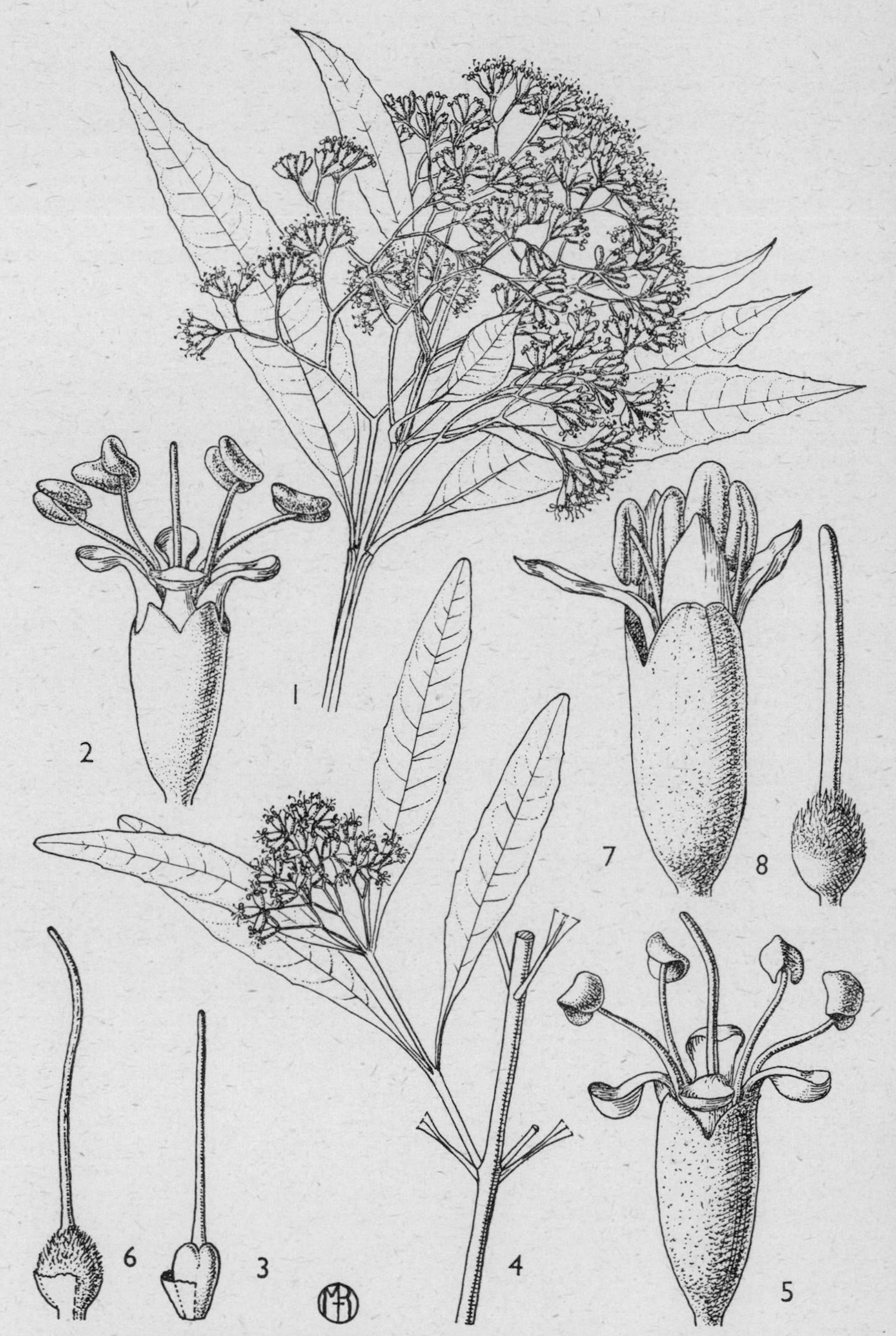

Fig. 8. *NUXIA FLORIBUNDA*—**1,** flowering branch, × $\frac{2}{3}$; **2,** flower, × 8 ; **3,** pistil, × 8 ; *N. OPPOSITIFOLIA*—**4,** flowering branch, × $\frac{2}{3}$; **5,** flower, × 8 ; **6,** pistil, × 8 ; *N. CONGESTA*—**7,** flower, × 8 ; **8,** pistil, × 8. 1–3, from *Drummond & Hemsley* 2797 4–6, from *Tweedie* 244.

1. **N. floribunda** *Benth.* in Hook., Comp. Bot. Mag. 2 : 59 (1836) & in DC., Prodr. 10 : 435 (1846) ; Fl. Cap. 4 (1) : 1039 (1909). Type : South Africa, Uitenhage District, on the Swartkops R., *Ecklon & Zeyher* 115 (K, holo. !)

Trees up to 20 m. tall or large shrubs ; bark pale brownish-grey. Lamina of leaves narrowly elliptic, up to 14 cm. long, 2–4 cm. wide, tapering above to a narrowly acute or acuminate apex, gradually convergent to a cuneate base, distantly denticulate to entire, glabrous. Inflorescences terminal and lax. puberulous when young, glabrate. Calyx 4 mm. long ; lobes cleanly dividing, Corolla white to yellow ; lobes rounded above, glabrous. Ovary glabrous. Fruit almost twice the length of the calyx, glabrous.

UGANDA. Kigezi District : Kachwekano Farm, Feb. 1950, *Purseglove* 3323 !
KENYA. Teita Hills, between Wuzi–Ngerenyi road and Bura bluff, Sept. 1953, *Drummond & Hemsley* 4392 !
TANGANYIKA. Kilimanjaro, Machame, Fau, Feb. 1928, *Haarer* 1027 ! ; Morogoro District : Uluguru Mts., Bunduki, on Mgeta R., Mar. 1953, *Semsei* 1095 ! ; Iringa District : Isata, July 1953, *Carmichael* 201 !
DISTR. **U**2 ; **K**7 ; **T**2–4, 6, 7 ; Belgian Congo, Nyasaland, Southern Rhodesia, South Africa.
HAB. Margins and relic patches of upland rain-forest ; 1200–2000 m.

SYN. *N. holstii* Gilg in P.O.A. C : 312 (1895) ; F.T.A. 4 (1) : 515 (1903). Type : Tanganyika, W. Usambara Mts., Mshusa, *Holst* 9138 (B, syn.†)
N. usambarensis Engl. in Abh. Preuss. Akad. Wiss. : 63 (1894) ; P.O.A. C : 312 (1895) ; Z.A.E. : 534 (1913). Type : Tanganyika, W. Usambara Mts., near Mshusa, *Holst* 9070 (B, holo.† ; BM, iso. !)
N. volkensii Gilg in P.O.A. C : 312 (1895). Type : Tanganyika, Kilimanjaro, *Volkens* 1686 (B, holo.†, K, iso. !)
N. polyantha Gilg in E.J. 30 : 376 (1901) ; F.T.A. 4 (1) : 513 (1903). Types : Tanganyika, Njombe District, Livingstone Mts., *Goetze* 1282 (B, syn.†, BM, isosyn. !) & Ukinga Mts., *Goetze* 988 (B, syn.†)
N. siebenlistii Gilg in Siebenlist, Forstw. Deutsch.-Ostafr. : 113 (1914) ; T.T.C.L.: 273 (1949). Type : Tanganyika, presumably a specimen from W. Usambara Mts. [perhaps *Siebenlist* 2331—see Fries in N.B.G.B. 8 : 698 (1924)]
Lachnopylis floribunda (Benth.) C.A. Sm. in K.B. 1930 : 25 (1930) ; I.T.U., ed. 2 : 166 (1952) ; T.T.C.L. : 270 (1949)
L. polyantha (Gilg) C.A. Sm. in K.B. 1930 : 18 (1930) ; T.T.C.L. : 271 (1949)

NOTE. This species is introduced into gardens in Nairobi district.

2. **N. oppositifolia** (*Hochst.*) *Benth.* in DC., Prodr. 10 : 435 (1846) ; Fl. Cap. 4 (1) : 1040 (1909). Type : Ethiopia, Tigré, *Schimper* 1714 (K, iso. !)

Trees or shrubs to 12 m. tall ; bark (reddish) brown. Lamina of leaves narrowly elliptic, up to 12 cm. long, 2·6 cm. wide, attenuating to a small rounded and often mucronulate apex, cuneate or narrowly cuneate at the base, bluntly serrate, ± sinuate or more rarely entire, glabrous. Inflorescences terminal and lateral, glabrous. Calyx 4–5 mm. long ; lobes cleanly dividing. Corolla-lobes rounded above, glabrous. Ovary hirsute, at least above. Fruit scarcely exceeding the calyx, hirsute.

UGANDA. Karamoja District : Kidipo River, *Brasnet* 141 !
KENYA. West Suk District : Kapenguria area, at foot of Suk escarpment, *Honore* 3716 ! ; Teita District : near Voi, Mbolo Hill, *Gardner* 3001 !
TANGANYIKA. Moshi District : Arusha Chini, on Pangani and Lumi Rivers, *Greenway* 8573 ! ; Mpanda District : Mahali Mts., Silambula, *Jefford & Newbould* 2072.
DISTR. **U**1 ; **K**1, 2, 7 ; **T**2, 4 ; Ethiopia, Sudan, Somaliland, Nyasaland, Northern and Southern Rhodesia, South Africa.
HAB. Riverine forest ; 750–2000 m.

SYN. *Lachnopylis oppositifolia* Hochst. in Flora 26 : 77 (1843) ; K.B. 1930 : 24 (1930) ; T.T.C.L. : 271 (1949) ; I.T.U., ed. 2 : 166 (1952) ; Fl. Pl. Sudan 2 : 381 (1952)
N. dentata Benth. in DC., Prodr. 10 : 435 (1846) ; P.O.A. C : 312 (1895) ; F.T.A. 4 (1) : 513 (1903) ; Fl. Cap. 4 (1) : 1040 (1909). Type : Ethiopia, unlocalized, *Salt* (BM, holo. !)

3. **N. congesta** *Fresen.* in Flora 21 : 606 (1838) ; P.O.A. C : 312 (1895) ; F.T.A. 4 (1) : 512 (1903) ; Z.A.E. : 534 (1913). Type : Ethiopia, unlocalized, *Rüppell* (FR, holo.)

Tree to 25 m. tall, bark dark grey-brown. Lamina of leaves either narrowly elliptic, up to 7 cm. long, 2 cm. wide, or elliptic to narrowly obovate, up to 11 cm. long, 5 cm. wide, convergent to an acute apex or rounded to a subtruncate and often mucronulate apex, cuneate below, entire, glabrous. Inflorescences terminal and usually congested, usually ± hairy. Calyx 5 mm. long ; lobes 5, often failing to separate cleanly (then apparently 3 or 4). Corolla-lobes deltoid above, usually with obscure hairs on the outer surface. Ovary longly hirsute, especially above. Fruit scarcely exceeding calyx, rounded above, densely hairy.

UGANDA. Acholi District : Imatong Mts., Lomwaga Mt., Apr. 1945, *Greenway & Hummel* 7282 ! ; Kigezi District : Soko Hill, *Purseglove* 800 ! ; Elgon, Buginyanya, *Snowden* 833.

KENYA. NE. Elgon, Jan. 1956, *Tweedie* 1369 ! ; Mau Forest, Sitoton camp, *Bally* 4732 ! ; Aberdare Mts., S. Kinangop, *D. C. Edwards* 2828/1 !

TANGANYIKA. Arusha District : Lubagai, *St. Clair-Thomson* 1251 ! ; Morogoro District : Uluguru Mts., Lupanga Peak, above Kibwe re-entrant, *Wigg* 937 ! ; Rungwe District : Kiwira River, *Greenway* 8386 !

DISTR. **U**1–3 ; **K**1, 3, 4, 6 ; **T**2–4, 6, 7 ; Belgian Congo, Sudan, Ethiopia, Nyasaland, Southern Rhodesia and probably farther to the west and south.

HAB. Upland rain-forest ; 1800–2700 m.

SYN. *Lachnopylis ternifolia* Hochst. in Flora 26 : 77 (1843). Type : Ethiopia, unlocalized, *Schimper*

Nuxia sambesina Gilg in P.O.A. C : 312 (1895) ; F.T.A. 4 (1) : 514 (1903). Type : Nyasaland, Zomba, *Kirk* (B, holo.†)

N. goetzeana Gilg in E.J. 30 : 375 (1901) ; F.T.A. 4 (1) : 514 (1903). Type : Tanganyika, Ukinga Mts., *Goetze* 1190 (BM, iso. !)

N. odorata Gilg in E.J. 30 : 376 (1901) ; F.T.A. 4 (1) : 513 (1903). Type : Tanganyika, Rungwe Mt., *Goetze* 1159 (BM, iso. !)

N. platyphylla Gilg in E.J. 32 : 141 (1902) ; F.T.A. 4 (1) : 512 (1903) ; Z.A.E. : 533 (1913). Type : Tanganyika, Kilimanjaro, Marangu, *Volkens* 1499 (BM, iso. !)

N. keniensis T.C.E. Fries in N.B.G.B. 8 : 697 (1924) ; T.T.C.L. : 273 (1949). Type : Kenya, western Mt. Kenya, *Fries* 777 (UPS, holo.)

N. latifolia T.C.E. Fries in N.B.G.B. 8 : 698 (1924). Type : Kenya, S. side Mt. Kenya, Kii River, *Fries* 2035 (UPS, holo., K, iso. !)

Lachnopylis congesta (Fresen.) C.A. Sm. in K.B. 1930 : 17 (1930) ; T.S.K. : 124 (1936) ; F.P.N.A. 2 : 59, t. 7 (1947) ; T.T.C.L. : 270 (1949) ; I.T.U., ed. 2 : 166 (1952) ; Fl. Pl. Sudan 2 : 381 (1952)

L. sambesina (Gilg) C.A. Sm. in K.B. 1930 : 17 (1930) ; T.T.C.L. : 271 (1949)

L. compacta C.A. Sm. in K.B. 1931 : 40 (1931) ; I.T.U., ed. 2 : 166 (1952) ; Fl. Pl. Sudan 2 : 381 (1952). Type : Sudan, Imatong Mts., above Kippia, *Chipp* 95 (K, holo. !)

L. platyphylla (Gilg) Dale in T.S.K. : 124 (1936) ; F.P.N.A. 2 : 60 (1947) ; T.T.C.L. : 271 (1949)

L. goetzeana (Gilg) Greenway in Burtt-Davy, Check-Lists of For. Trees & Shrubs of Brit. Emp. No. 5 (1) : 112 (1940) ; T.T.C.L. : 271 (1949)

L. odorata (Gilg) Greenway in Burtt-Davy, Check-Lists of For. Trees & Shrubs of Brit. Emp. No. 5 (1) : 112 (1940) ; T.T.C.L. : 271 (1949)

NOTE. An aggregate species whose range may possibly be even further increased by the inclusion of other synonyms. *N. mannii* Gilg from British Cameroons and *N. angolensis* Gilg from Angola are scarcely specifically distinct, and others (e.g. *N. dekindtii* Gilg and *N. rupicola* Gilg) are very closely related. In the very widest sense our species would include *N. emarginata* Sond. and its allies, which would extend its range to South Africa.

VARIATION. The variation of this species is well worthy of detailed field study. A general feature is the occurrence of a clinal pattern in two characters. The distribution of the species follows the line of the mountain masses on the margins of the Rift Valley : from the north and west limit in Uganda it runs in a roughly semicircular band passing between Mt. Kenya and the Mau Escarpment, on through the volcanic peaks of north-east Tanganyika to the mountains north of Lake Nyasa. Within this range the leaf-breadth increases from north to south ; leaves three or more times as long as broad are found only in Uganda and, to a lesser extent, in western Kenya. The division of the calyx irregularly into lobes is common in Tanganyika, much less so in Kenya, while in Uganda regular 5-lobed calyces are usual, especially in the north.

Extreme variation in other characters combines with those just mentioned to give distinct appearances to some local groups and this has led to some of them being described as species. They may be referred to as races and some examples are :—

The Mt. Kenya race (" *N. keniensis* ")—with the inflorescences much laxer than usual and the indumentum of the calyx at its densest.

The Mt. Hanang race—with almost glabrous calyces and leaves of a characteristic rhomboid shape.

The Mt. Debasien race—with the outer face of the corolla glabrous, the leaves narrow and the calyx regularly divided.

The genus requires revision as a whole. This may lead to fewer species being accepted than the sixty or more that have been described, and consequently to the delimitation of infraspecific groups for which then our rather vague concept of race may no longer be suitable.

Excluded genus

Gaertnera *Lam.*, Illustr. 2 : 273 (1819) ; G.P. 2 : 798 (1876) ; F.T.A. 4 (1) : 543 (1903) ; T.T.C.L. : 270 (1949)

This genus will be included in this Flora in the family *Rubiaceae*, with which it is more consistent in habit, anatomy and the details of its ovary (see Solereder in E. & P. Pf. 4 (2) : 27 (1892)).

INDEX TO LOGANIACEAE